HARNESSING THE FOURTH INDUSTRIAL REVOLUTION THROUGH SKILLS DEVELOPMENT IN HIGH-GROWTH INDUSTRIES IN CENTRAL AND WEST ASIA

UZBEKISTAN

MAY 2023

ADB

ASIAN DEVELOPMENT BANK

© 2023 Asian Development Bank
6 ADB Avenue, Mandaluyong City, 1550 Metro Manila, Philippines
Tel +63 2 8632 4444; Fax +63 2 8636 2444
www.adb.org

Some rights reserved. Published in 2023.

ISBN 978-92-9270-137-6 (print); 978-92-9270-138-3 (electronic); 978-92-9270-139-0 (ebook)
Publication Stock No. TCS230156-3
DOI http://dx.doi.org/10.22617/TCS230156-3

Notes:
In this publication, "A$" refers to Australian dollars, "₹" refers to Indian rupees, and "$" refers to United States dollars. ADB recognizes "Vietnam" as Viet Nam and "Korean" as referring to the Republic of Korea.

Notes on names of ministries of the Government of Uzbekistan: This publication refers to ministry names current when the content was completed. Changes were later announced in the Decree of the President of the Republic of Uzbekistan No. 269, *On measures to implement the administrative reforms of New Uzbekistan* (21 December 2022), to take effect in January 2023. The new ministries under this restructuring related to the discussion in this publication include the Ministry of Employment and Poverty Reduction and the Ministry of Higher Education, Science and Innovation. Other government agencies referred to in the publication may also have been affected.

Cover design by Cleone Baradas

Printed on recycled paper

Contents

Tables, Figures, and Boxes ... v

Foreword ... ix

Preface and Acknowledgments ... xi

Abbreviations ... xiii

Executive Summary ... xiv

Scope and Methodology of the Study ... xiv

Key Findings for Uzbekistan ... xv

Key Recommendations and Way Forward ... xviii

Textile and Garment Manufacturing Industry ... xx

Construction Industry ... xx

1   The Industry 4.0 Skills Challenge ... 1

A.   Industry 4.0 and Its Relevance for Uzbekistan ... 1

B.   Industry Selection ... 3

C.   Textile and Garment Manufacturing ... 4

Relevance of Industry 4.0 to the Textile and Garment Industry ... 4

Skills Demand Analysis ... 10

Skills Supply Trends ... 19

D.   Construction Industry ... 21

Relevance of Industry 4.0 to the Construction Industry ... 22

Skills Demand Analysis ... 27

Skills Supply Trends ... 33

E.   Emerging Jobs ... 36

2   Overview of the Training Landscape ... 38

A.   Industry 4.0 Readiness and Impact of COVID-19 ... 39

B.   Curricula ... 40

C.   Industry Engagement ... 43

D.   Teachers, Trainers, and Instructors ... 45

E.   Performance and Policy Support ... 46

F.   Supply and Demand Mismatches ... 48

3   National Policy Responses   **51**

A.   Overview of Industry 4.0 Policy Landscape   51

B.   Assessment of Current Policy Approaches in Uzbekistan Related to Industry 4.0 and Skills   55

Assessment of Policy Actions ("The What")   55
Assessment of Implementation of 4IR Policies ("The How")   57

C.   Assessment of 4IR Policies in Relation to the COVID-19 Pandemic   59

4   The Way Forward   **60**

A.   Summary of Key Challenges to Industry 4.0 Adoption Faced by Uzbekistan   60

B.   Recommendations to Address Challenges   61

Recommendation 1: Develop sectoral 4IR adoption plans to coordinate technology adoption and skills development   63

Recommendation 2: Develop programs to guide digital transformation of small and medium businesses   64

Recommendation 3: Strengthen the training capabilities of employers   65
Recommendation 4: Develop online learning platforms   67
Recommendation 5: Adopt programs to strengthen industry knowledge of trainers   67
Recommendation 6: Promote the use of innovative technologies to strengthen training delivery   68
Recommendation 7: Develop targeted programs to ensure that women can benefit from 4IR   70
Recommendation 8: Develop skilling and labor support programs for digital freelancers   70

C.   Industry-Specific Priorities   72

Textile and Garment Manufacturing Industry   72
Construction Industry   73

Appendix: Participants in the National Consultations   **74**

References   **77**

Tables, Figures, and Boxes

Tables

1	Primary Data Sources Used	2
2	Occupational Groups in the Textile and Garment Manufacturing Industry	13
3	Categories of Skills Considered in the Analysis	17
4	Four Types of Training Channels	20
5	Occupational Groups in the Construction Industry	28
6	Key Policies Relevant to Managing the Impact of Industry 4.0 on Skills in Uzbekistan	53
7	Recap of Challenges Facing Uzbekistan in Relation to Industry 4.0	60
8	Summary of Recommendations, Potential Lead Agencies, and Approximate Time Frame for Implementation	62
9	Summary of Findings in the Textile and Garment Manufacturing Industry	72
10	Summary of Findings in the Construction Industry	73
A1	Stakeholders Engaged in Initial Consultations for Uzbekistan	74
A2	Stakeholders Engaged in Further Consultations for Uzbekistan	75

Figures

1	Employers' Understanding of Industry 4.0 Technologies in the Textile and Garment Manufacturing Industry in Uzbekistan	6
2	Employers' Understanding of Industry 4.0 Technologies among Firms in the Textile and Garment Manufacturing Supply Chain in Uzbekistan	7
3	Expected Increase in Output Per Worker Due to Industry 4.0 Technologies in the Textile and Garment Manufacturing Industry in Uzbekistan, 2020–2025	8
4	Current and Future Adoption of Relevant Industry 4.0 Technologies in the Textile and Garment Manufacturing Industry in Uzbekistan	9
5	Perceptions on Impact of the COVID-19 Pandemic on Adoption of Industry 4.0 Technologies in the Textile and Garment Manufacturing Industry in Uzbekistan	10
6	Estimated Impact of Industry 4.0 on Number of Jobs in the Textile and Garment Manufacturing Industry in Uzbekistan, 2020–2025	11
7	Employers' Expectations on Impact of Industry 4.0 on the Number of Jobs in the Textile and Garment Manufacturing Industry in Uzbekistan, 2020–2025	13
8	Composition of Jobs in 2020 and by 2025 in the Textile and Garment Manufacturing Industry in Uzbekistan	14

9   Estimated Net Job Gains by Gender from Industry 4.0 Adoption in the Textile and Garment   15
     Manufacturing Industry in Uzbekistan, 2020–2025

10   Time Spent by Employees on Tasks at Work in 2020 and by 2025 in the Textile and Garment   16
     Manufacturing Industry in Uzbekistan

11   Importance of Skills in 2020 and for Industry 4.0 Adoption by 2025 in the Textile and Garment   18
     Manufacturing Industry in Uzbekistan

12   Required Step-Up in Level of Proficiency of Employees' Skills from 2020 for Industry 4.0 Adoption   18
     by 2025 in the Textile and Garment Manufacturing Industry in Uzbekistan

13   Employer Sentiment Toward Graduates Hired in the Textile and Garment Manufacturing Industry   19
     in Uzbekistan

14   Employers' Perception on Training for Employees in the Textile and Garment Manufacturing Industry   20
     in Uzbekistan

15   Proportion of Employees Receiving Training in 2020 and Requiring Training by 2025 in Each Training   21
     Channel in the Textile and Garment Manufacturing Industry in Uzbekistan

16   Understanding of Industry 4.0 Technologies in the Construction Industry in Uzbekistan   23

17   Understanding of Industry 4.0 Technologies Among Firms in the Construction Supply Chain   24
     in Uzbekistan

18   Expected Increase in Output per Worker Due to Industry 4.0 Technologies in the Construction Industry   25
     in Uzbekistan, 2020–2025

19   Current and Future Adoption of Relevant Industry 4.0 Technologies in the Construction Industry   26
     in Uzbekistan

20   Perception on the Impact of the COVID-19 Pandemic on the Adoption of Industry 4.0 Technologies   27
     in the Construction Industry in Uzbekistan

21   Estimated impact of Industry 4.0 on Number of Jobs in the Construction Industry in Uzbekistan,   28
     2020–2025

22   Expected Impact of Industry 4.0 on the Number of Jobs between 2020 and 2025 in the Construction   30
     Industry in Uzbekistan

23   Composition of Jobs in 2020 and by 2025 by Occupational Group in the Construction Industry   31
     in Uzbekistan

24   Estimated Net Job Gains by Gender from Industry 4.0 Adoption Between 2020 and 2025   31
     in the Construction Industry in Uzbekistan

25   Time Spent by Employees on Tasks at Work in 2020 and by 2025 in the Construction Industry   32
     in Uzbekistan

26   Importance of Skills in 2020 and for Industry 4.0 Adoption by 2025 in the Construction Industry   33
     in Uzbekistan

27   Required Step-Up in Employee Proficiency Level from Today for Industry 4.0 Adoption   34
     in the Construction Industry in Uzbekistan, 2020–2025

28   Employer Sentiment Toward Graduates Hired in the Construction Industry in Uzbekistan   34

29   Employers' Perception on Training for Employees in the Construction Industry in Uzbekistan   35

30   Proportion of Employees Receiving Training in 2020 and Requiring Training by 2025   36
     for the Construction Industry in Uzbekistan

31   Job Roles Expected to Become More Prominent with the Adoption of Industry 4.0 Technologies   37
     between 2020 and 2025

32   Perception of Training Institutions on Readiness for Industry 4.0 in Uzbekistan   39

33 Impact of COVID-19 on Training Institutions in Uzbekistan   40
34 Frequency of Review and Update of Curricula by Training Institutions in Uzbekistan   41
35 Prevalence of Industry 4.0-Related Courses and Industry 4.0-Based Delivery in Training Institutions in Uzbekistan   41
36 Current Adoption of Specific Industry 4.0 Technologies by Employers and Prevalence of Courses Relevant to These Technologies in Training Institutions in Uzbekistan   42
37 Partnership Activities Between Training Institutions and Employers in Uzbekistan   43
38 Partnership Activities Between Employers and Training Institutions in Uzbekistan   44
39 Frequency of Communication Between Employers and Training Institutions in Uzbekistan   45
40 Training Institutions' Practices to Support Instructors in Uzbekistan   46
41 Training Institutions' Perceptions on and Reasons for Difficulty in Filling Places in Uzbekistan   47
42 Training Institutions' Perceptions on Most Impactful Public Policies for Training Provision in Uzbekistan   47
43 Training Institutions' Perception of Reasons for Students Being Unable to Find Jobs upon Graduation in Uzbekistan   48
44 Non-Training Initiatives Provided by Training Institutions to Support Trainees in Their Professional and Personal Development in Uzbekistan   49
45 Perception of Employers on Graduates' Preparedness for Entry-Level Positions in Uzbekistan   50
46 Degree of Focus of Policy Actions to Manage the Impact of Industry 4.0 on Jobs and Skills in Uzbekistan   55
47 Implementation Challenges Associated with Industry 4.0 Policies for Jobs and Skills in Uzbekistan   58
48 Relevant Best Practices That Could be Adopted to Tackle Challenges in Adoption of Industry 4.0 Practices   62

Boxes
1 Estimating Employment Changes Due to Adoption of Industry 4.0 Technologies   12
2 The Potential Cost of Government Inaction on Job Creation under Industry 4.0 Adoption   29
3 Helping Small and Medium-Sized Enterprises Go Digital in Australia and Singapore   65
4 Programs to Strengthen Employers' Training Capabilities in Ireland, Malaysia, Pakistan, and Singapore   66
5 Programs to Strengthen Industry Knowledge of Trainers in Malaysia and the United Kingdom   68
6 Use of Artificial Intelligence, Virtual Reality, and Augmented Reality Technologies to Improve Learners' Experience in India, South Africa, and the United States   69
7 Various Forms of Government Support for Digital Freelancers in Pakistan   71

Foreword

The spectacular spread of Fourth Industrial Revolution (4IR) technologies globally has brought great upheavals and disruptions in labor markets. The *2020 Future of Jobs* report by the World Economic Forum estimated that by 2025, 85 million jobs may be displaced by a shift in the division of labor between humans and machines, while 97 million new roles could emerge that address the new realities of division of labor between humans, machines, and algorithms. The erstwhile fears of major job losses have given way to a more balanced discourse on amplifying the promise of technologies for sustainable development and human well-being and minimizing the peril of unemployment. While the new generation of disruptive technologies each have their own unique strength, it is the collective potential of these technologies to improve productivity and the quality of goods and services that have the greatest promise of influencing societal value and impact. Fusing the boundaries between the physical, digital, and biological worlds, the 4IR technologies that include artificial intelligence, robotics, the Internet of Things, 3D printing, genetic engineering, quantum computing, and machine learning, are fast becoming indispensable to modern work life, and indeed to the daily lives of citizens. The question is no longer how to prepare for 4IR technologies tomorrow but how to help individuals, firms, and societies today to effectively draw on them for greater productivity and prosperity.

To provide insights on the impact of 4IR on skills and jobs, the Asian Development Bank undertook the study "Harnessing the Fourth Industrial Revolution Through Skills Development in High-Growth Industries in Central and West Asia." The study suggests concrete pathways by which developing member countries can shape the transition of their economies to 4IR technologies to tap the potential for productivity and new jobs. It provides insights on opportunities, challenges, and promising approaches in 4IR for Azerbaijan, Pakistan, and Uzbekistan with specific focus on two industries in each country deemed important for growth, employment, and 4IR—transportation and storage and agro processing in Azerbaijan, information technology-business process outsourcing and textiles and garment manufacturing in Pakistan, and construction and textiles and garment manufacturing in Uzbekistan. The Central and West Asia region can benefit tremendously by tapping into 4IR technologies. It is important for the region to effectively manage the transition to 4IR technologies for greater economic diversification, moving up the global value chain and strengthening knowledge-based growth processes.

A key aspect of embracing 4IR technologies is to invest in appropriate skills. Based on a recent study by Amazon and Workplace Intelligence, 78% of Gen Z and millennial employees are concerned they lack the skills required to advance their career, and 58% are afraid that their skills have gone stale since the onset of the pandemic, and as many as 70% feel unprepared for the future of work. Hence the time to act on skills development is now, with ever-increasing demand for skills. The coronavirus disease (COVID-19) pandemic has caused incursions in business processes that have led to the acceleration of digital solutions in the marketplace. With the digital talent gap growing, there is a need for both public and private sector entities to invest in re-skilling and upskilling for new and transforming jobs due to adoption of technologies. The study stresses the importance of on-the-job training for 4IR technologies and the need for governments to embark on deliberate strategies for life-long-learning opportunities.

The study affirms a positive outlook to 4IR creating new opportunities for quality jobs. While many jobs will indeed be lost as a result of automation, new jobs will emerge through the adoption of technologies that will increase worker productivity and competitiveness of nations, thereby leading to greater prosperity. However, tapping such benefits is predicated on increasing investments in skills development and greater efforts by companies to upskill their workforce to perform new and higher order roles in complementarity with machines. The study has resulted in a suite of country reports for Azerbaijan, Pakistan, and Uzbekistan and a synthesis report that captures common elements across the three. The reports provide policy makers with evidence-based solutions for skills and talent development to strengthen the countries' readiness for a transition to 4IR.

The study highlights that while job losses will be real, a well-prepared 4IR strategy with industry transformation road maps that are recommended in the study can convert disruptions to opportunities to pivot the workforce to new and modern occupations. In light of post-COVID-19 realities, digital transformation and technology adoption can make enterprises more agile and responsive to changing market conditions.

We believe that 4IR technologies can not only bring greater economic value to enterprises and individuals, they can also help to strengthen the pathways for sustainable and inclusive development. There is more work to be done to explore and leverage the benefits at the intersection of digitalization and climate resilience and to scale up the deployment of 4IR technologies for equity and increasing opportunities for vulnerable populations. We welcome ideas and partnerships with stakeholders as we pursue this area of research toward concrete implementation and next level of analytical work.

Bruno Carrasco
Director General
Sustainable Development and
Climate Change Department
Asian Development Bank

Yevgeniy Zhukov
Director General
Central and West Asia
Department
Asian Development Bank

Preface and Acknowledgments

The Asian Development Bank (ADB) study "Harnessing the Fourth Industrial Revolution Through Skills Development in High-Growth Industries in Central and West Asia" addresses a crucial topic of great relevance to labor markets and jobs. At the heart of this study is the quest to better understand how disruptive technologies are influencing the nature of jobs and skills. Technologies of the Fourth Industrial Revolution (4IR) are influencing every sector and sphere of economies and societies, whether manufacturing or services. At the same time, business processes such as marketing, storage, transport, logistics, and payment mechanisms are greatly transformed with digital technologies. Business practices have been disrupted and reengineered through frontier technologies such as artificial intelligence, digital twins, robotics, and 3D printing.

We bring this piece of research to inform policy makers and practitioners of the implications of 4IR for future job markets. The study drew on various sources of secondary and primary data. It included surveys of employers and training institutions to assess their readiness for 4IR. The study presents analysis of data from online job portals from each of the countries covered in the study – Azerbaijan, Pakistan, and Uzbekistan—to assess trends in skills demand.

The study used a modeling exercise to estimate both job displacement and gains in select industries in the 3 countries. A review of the policy landscape based on benchmarks from international experiences provides the basis for the action points that developing countries can use to harness the potential of Industry 4.0 to increase productivity, facilitate skills development, and incentivize industry. The findings and recommendations from the study underscore the need for renewing skills development strategies with a full life cycle approach. This means that there are no degrees or certificates for life and regular upskilling is essential. The preponderant focus on institution-based training needs to give way to more flexible and multimodal training to include bootcamps, e-learning, and work-place based training. Training for digital skills at basic, intermediate, and higher levels needs a significant ramp up as workplaces undergo digital transformation. The benefits of 4IR can only be effectively harnessed if adequate investments are made in skills development.

The study was led by Shanti Jagannathan, in partnership with Eisuke Tajima and ADB team members. Rie Hiraoka and Brajesh Panth provided valuable guidance to the study. We thank the consultant team led by Fraser Thompson, director, AlphaBeta, for an excellent partnership in this study, together with Wan Ling Koh and Shivin Kohli. AlphaBeta's team developed the analytical model for the study and collaborated closely with ADB's team to bring new insights and directions and we are grateful for this professional collaboration. We thank Xin Long, Aziz Haydarov and Kevin Corbin from ADB headquarters and representatives of ADB resident missions in Azerbaijan, Pakistan, and Uzbekistan, respectively, for their valuable support and country-level consultations (Sabina Jafarova, Sanan Shabanov, Yuliya Hagverdiyeva and Elvin Imanov from Azerbaijan; Khuram Imtiaz and Rizwan Haider from Pakistan; and Farida Djumabaeva and Shahina Rismetova from Uzbekistan). Joehanne Kristal Santos and Evangelyn Medina from ADB provided timely coordination of meetings and activities during the study. Cherry Zafaralla copy edited this report. Dorothy Geronimo coordinated the editorial and publication process with ADB consultants: Maria Theresa Mercado (proofreading), Mariel Gabriel (proof checking), and Edith Creus (typesetting), and Cleone Baradas (cover design).

The study benefited greatly from enriching discussions with government representatives in the respective countries. Early workshops with government representatives and experts were held to inform the study process. The findings of the study were shared in country level workshops. Senior officials and key counterparts consulted are listed at the end of each country report. Tamerlan Tagiyev, Head, Center for Analysis and Coordination of the Fourth Industrial Revolution (Azerbaijan); Shabnum Sarfraz, member, Social Sector and Devolution, Planning Commission, Ministry of Planning, Development and Special Initiatives, Asadullah Faiz, member, Private Sector Development, Punjab Planning and Development Board, and Salman Shami, member, Private Sector Development, and Muhammad Haroon Naseer, additional director general, Punjab Skill Development Authority (Pakistan); and Oybek Shagazatov, head, Main Department of Cooperation with International Financial Organisations, Ministry of Investments and Foreign Trade (Uzbekistan). Several experts also contributed to the study—Amin Charkazov, Ramil Azmammadov (agro-processing) and Sabuhi Abdurahmanov (transport) from Azerbaijan; Allah Bakhsh Malik, Nasir Amin, Muhammad Asim Rehmat (information technology-business process outsourcing) and Muhammad Babar Ramzan (textiles) from Pakistan; and Shukhrathoja Amanov, Khabibullaev Shavkat Azamatovich (construction), and Umida Vakhidova (textile) from Uzbekistan.

We look forward to discussions in taking forward the study's policy recommendations.

Sungsup Ra
Chief Sector Officer
Sustainable Development and
Climate Change Department
Asian Development Bank

Abbreviations

4IR	Fourth Industrial Revolution (or Industry 4.0)
ADB	Asian Development Bank
AR/VR	augmented reality and/or virtual reality
BAU	business-as-usual
COVID-19	coronavirus disease
ICT	information and communication technology
ILO	International Labour Organization
IT–BPO	information technology–business process outsourcing
IOT	Internet of Things
ITM	industry transformation map
SMEs	small and medium-sized enterprises
STEM	science, technology, engineering, and mathematics
TVET	technical and vocational education and training

Executive Summary

The coronavirus disease (COVID-19) pandemic is accelerating the digital transformation of businesses and jobs across all industries. The Asian Development Bank (ADB) Asian Economic Integration Report 2021 found that accelerated digital transformation can potentially boost global output, trade and commerce, and employment, with Asia expected to reap an economic dividend of more than $1.7 trillion yearly (equivalent to 6.1% of the 2020 regional gross domestic product baseline), or more than $8.6 trillion over the 5-year projection of the study to 2025.

Against this climate, the influence of disruptive technologies on jobs and labor markets has intensified worries around extensive job losses arising from automation and potential disappearance of the comparative advantage of countries based on competitive labor costs. The readiness of developing countries to effectively address the transition to the Fourth Industrial Revolution (4IR) or Industry 4.0 has become an important policy concern. To better understand the implications of 4IR on the future of jobs and to assess the readiness of education and training institutions to prepare workers for future jobs, ADB undertook this study that seeks to capture the anticipated transformations on jobs, tasks, and skills; and to outline policy directions to prepare the workforce for future jobs, particularly in the post-COVID-19 world.

Scope and Methodology of the Study

The study comprises four volumes or reports covering three countries in the Central and West Asia region: Azerbaijan, Pakistan (with focus on Punjab), and Uzbekistan; including a synthesis report that draws together the findings from the three country studies. This report on Uzbekistan is the third of the four volumes, while the report on Azerbaijan is volume 1 and volume 2 is the report on Pakistan. The synthesis report outlines common areas of policy and action for Industry 4.0.

The study has the following features:

(i)   Two focus industries were selected in each country that are crucial for growth, employment, and 4IR, namely, agro-processing and transportation and storage in Azerbaijan; textile and garment manufacturing and information technology–business process outsourcing (IT–BPO) in Pakistan; and textile and garment manufacturing and construction in Uzbekistan.

For the focus industries, a survey of employers and training institutions, and analysis of data from online job portals to assess trends in skills demand and supply were conducted. The data collected is used to estimate job displacement and gains in the selected industries in each country through a logical model based on economic principles governing job creation and displacement.

(ii)   The policy landscape is also assessed. To understand gaps in the policy landscape in harnessing the potential of 4IR to increase productivity and create quality jobs, the study considered the comprehensiveness of policies in terms of stimulating 4IR adoption and worker reskilling efforts, creating new flexible qualification pathways, and building inclusiveness to extend the benefits of 4IR to all workers. The strength of implementation of policies, particularly against the backdrop of the COVID-19 pandemic, was also assessed.

Surveys of employers and stakeholders were conducted between June and September 2021, and country data from 2020 (the latest for which full-year data is available) was used. To align all baselines for comparison, the survey data collected is assumed to be reflective of perspectives and circumstances as of end of 2020. The objective is to provide an illustrative view of how 4IR can impact jobs and skills in the three countries in the Central and West Asia region in a 5-year period between 2020 and 2025.

Key Findings for Uzbekistan

This report covers the key findings for Uzbekistan. The report analyzed the implications of 4IR for jobs, tasks, and skills in the textile and garment manufacturing industry, as well as the construction industry. These industries were selected based on several factors including their importance to national employment and growth prospects, the degree of relevance of 4IR technologies to the industry, and alignment with national growth plans.

The textile and garment manufacturing industry has been identified in national growth plans such as the Five-Area Development Strategy, 2017–2021 as an industry with significant untapped potential to drive long-term economic growth in Uzbekistan. The country's recent admission to the special incentive arrangement for sustainable development and good governance under the unilateral Generalized Scheme of Preferences scheme by the European Union will increase the competitiveness of Uzbekistan's textile products in Europe and provide further growth impetus for the industry. Past research demonstrates the significant potential for 4IR technologies to be adopted in the textile and garment manufacturing industry, with productivity gains of up to 46%.

For the construction industry, it is already one of the largest employers in Uzbekistan, contributing 9.2% to national employment in 2018, and features strongly in national growth plans. The Strategy of Modernization and Innovative Development of the Construction Industry, 2021–2025 sets out plans for the adoption of innovative technologies, and human capital development. As in the textile and garment manufacturing industry, the adoption of 4IR technologies could bring significant benefit to the construction industry, with past research estimating that full-scale digitization could generate an estimated 12%–20% in annual cost savings. It is therefore timely for policy makers to consider how 4IR technologies can be integrated into both industries' growth strategies.

The report finds that 4IR will have a transformational effect on jobs and skills in the textile and garment manufacturing, and construction industries in Uzbekistan with strong potential for positive gains in jobs and productivity, which can be reaped through adequate investments in skills and training.

Key findings from the report include the following:

(i)   **Industry 4.0 technologies are perceived by firms to have the potential to transform productivity in their industries, but many firms, particularly in the construction sector, lack understanding of these technologies.**

(a)   Some 29% of textile and garment manufacturing firms and 51% of construction firms surveyed currently have a limited understanding of 4IR technologies and their applications and would need further support to adopt such technologies.

(b)   Most firms surveyed indicated that they are keen to deploy 4IR technologies across a range of function by 2025 with Internet of Things technologies prioritized by firms in both industries. There are strong expectations regarding potential gains from adopting 4IR technologies. Textile and garment manufacturing firms estimate that the adoption of 4IR technologies will increase labor productivity (i.e., output per worker) by 68% in between 2020 and 2025 while construction firms estimate a 60% increase.

(ii)   **The full adoption of 4IR technologies could bring net job gains in the textile and garment manufacturing and construction industries in Uzbekistan.**

(a)   The study estimates that the adoption of 4IR technologies in the textile and garment manufacturing and construction industries will create net job gains by comparing the displacement and productivity effects of adopting 4IR technologies. The number of new jobs created by productivity gains from adopting 4IR technologies will exceed the number of jobs displaced by automation.

(b)   As a result of 4IR technologies adoption, more jobs are expected to be created—100,000 new jobs in Uzbekistan's textile and garment manufacturing industry and 334,000 new jobs in the construction industry. This is over and beyond the business-as-usual job growth in a scenario without 4IR adoption, meaning that these jobs are fully attributable to 4IR adoption. These jobs gains can only be realized if policymakers adopt policies to support the adoption of 4IR technologies in firms as well as build a 4IR-ready workforce able to support the industry's transformation.

(iii)   **While manual roles are expected to remain a key component of both industries, the tasks undertaken by workers and skills required will change.**

(a)   Employer surveys show that even with the adoption of 4IR technologies, employers expect that a significant proportion of jobs in 2025 will continue to be manual jobs. More than half of all jobs in the textile and garment manufacturing industry in 2025 are expected to be manual jobs while a third of construction jobs are expected to be in manual roles. In the textile and garment manufacturing industry, this is likely due to firms preferring to maintain handmade elements in their products as well as overall expectations that the industry will grow rapidly and increase demand for workers in all roles. In the construction industry, this can be attributed to the technical challenges in automating construction jobs fully.

(b)   Notwithstanding the high proportion of manual jobs that will remain, the tasks taken by workers within these roles could change and new skills could be required. The employer surveys show that employers expect workers to spend less time on routine, physical tasks in an average work week and more time on nonroutine tasks and analytical tasks. This could mean that manual workers will increasingly be expected to operate basic machines or software and continual reskilling would be needed to ensure that they remain relevant as the skills requirements for manual work changes.

(iv)   **Stronger alignment on future skills needs is needed between employers and training institutions.**

(a)   The adoption of 4IR technologies will change the skill needs of employers in both industries by 2025. For textile and garment manufacturing and construction firms, digital and/or information and communication technology (ICT) skills as well as creative thinking and design skills will become the most valued skills by 2025, and it is critical that training institutions are equipped to prepare workers for these skill changes.

(b) Existing gaps in graduate quality would need to be bridged and the frequency of communication between training institutions and employers further increased. While 79% of training institutions assess their graduates to be adequately prepared for entry-level positions, only 47% of textile and garment manufacturers and 65% of construction firms take the same view. Close to half of construction companies and a quarter of textile and garment manufacturing companies reported that they communicate with training institutions less than once a year or never communicate with training institutions. Employers currently need to provide a significant amount of training to their workers to ensure that they can do their jobs well, with employers in both industries providing on-the-job training (OJT) to around 70% of employees as of 2020.

(v) **Training institutions require additional financial and technical support to adopt innovative training approaches and be 4IR-ready.**

(a) Among the training institutions surveyed, 47% strongly agree that additional technical and financial support is needed to enable them to prepare workers for 4IR. As of 2020, 73% of training institutions surveyed offer general digital skills programs to improve digital literacy and over 50% use technology-based approaches such as online self-learning modules and interactive videos. There is scope to further strengthen the effectiveness of training delivery by implementing 4IR-enabled teaching approaches in the classroom to ensure that graduates are familiar with the use of advanced technologies.

(b) Strong collaboration with industry is also critical to enable training institutions to create a 4IR-ready workforce. While around 60% of employers surveyed in both industries have adopted IOT technologies, only 30% of training institutions offer courses relevant to IOT. Industry collaboration can help to relieve the possible resource constraints faced by training institutions (e.g., the lack of equipment or qualified instructors) while ensuring that training programs are aligned with industry's needs.

(vi) **There is a need to strengthen coordination on future skills needs among government, industry, and training institutions to ensure that Uzbekistan can reap the gains of 4IR.**

(a) Uzbekistan has adopted a range of policies to enable firms and workers to transform digitally but lacks a clear 4IR road map to coordinate these policies. Plans to encourage innovation among firms in Uzbekistan are set out in national strategies, such as the Digital Uzbekistan 2030 road map and Strategy for Innovative Development 2019–2021. There are measures to support firms in specific industries set out in various decrees. In parallel, there are efforts to make training more relevant to industry needs, with sectoral skills councils established to update professional qualifications standards.

(b) Despite these efforts, the quality and depth of coordination will need to be strengthened among training institutions, employers, and government going forward. The training institution and employer surveys highlighted several mismatches in terms of views on quality of graduates and the relevance of training courses to industry's needs, which points to a lack of alignment in skill needs across industry and training institutions. Stronger policy focus is also needed to build awareness of 4IR technologies among small and medium-sized enterprises; and develop targeted programs to ensure that women can also benefit from 4IR.

Key Recommendations and Way Forward

Drawing on the findings of the industry and training institution surveys as well as the policy assessment, the report identifies eight recommendations for Uzbekistan to strengthen its preparedness toward 4IR. These include recommendations to strengthen both the demand for skills (i.e., creation of 4IR-related jobs) and the supply of skills.

The following list provides a summary of recommendations and potential lead agencies in Uzbekistan to implement each recommendation as well as the approximate implementation timeframe.

The specific actions for each of the eight Uzbekistan recommendations include the following:

(i) **Develop sectoral 4IR adoption plans to coordinate efforts in promoting technology adoption and skills development.** Uzbekistan has adopted a range of policies to enable firms and workers to transform digitally but has scope to strengthen coordination in implementing these policies. A clear action plan for 4IR adoption in each industry could strengthen coordination among stakeholders and facilitate long-term planning for firms and training institutions. Uzbekistan could consider the development of 4IR action plans modelled after Singapore's Industry Transformation Maps, which provide information on technology impacts, labor market shifts, the skills required for different occupations and reskilling options for different industries. The Ministry for Innovative Development could take the lead in coordinating these plans and this could be implemented over the next 12 to 36 months.

(ii) **Develop programs to guide digital transformation of small and medium businesses.** Understanding and adoption of 4IR technologies vary across firms in Uzbekistan, with SMEs expected to face more resource constraints in adopting such technologies. Programs to guide the digital transformation of firms can ensure that SMEs also benefit from 4IR. In Australia, the Digital Solutions – Australian Small Business Advisory Services program provides subsidized independent advice to Australian small businesses to help them build their digital capabilities. In Uzbekistan, the Agency for Development of Small Business and Entrepreneurship and Ministry for Development of Information Technologies and Communications could lead efforts to identify the specific challenges faced by SMEs in adopting new technologies and design programs to counter these challenges. This could be implemented over the next 12 to 36 months.

(iii) **Strengthen the training capabilities of employers.** A large proportion of employers in Uzbekistan provide OJT to their employees. Programs to support employers to strengthen their training capabilities—both in-house and through training providers—can help to strengthen the effectiveness of the training provided. In Pakistan, the Punjab Skills Development Fund's (PSDF) Industry Training Programs provide support to businesses to train youths, to ensure that the training curriculum is agile and in line with the emerging skills needs of businesses. The PSDF funds the training and pays trainees monthly stipends to incentivize them to complete the courses but requires a dedicated classroom and instructor to ensure that proper training is carried out. Such programs could be implemented through agencies such as the Ministry of Higher Education in Uzbekistan, Ministry of Employment and Labor Relations in the next 12 to 36 months.

(iv) **Develop online learning platforms.** In Uzbekistan, a National Qualifications Framework has been established and there are plans to establish competency assessment centers to recognize the results of nonformal learning and award professional qualifications to workers. However, opportunities for the recognition of qualifications acquired outside formal education (e.g., short-term training courses offered by online providers) remains limited. Uzbekistan could consider online learning platforms to rapidly build up the new skills required by employers. For instance, in the Republic of Korea, online learning platforms have been established to upskill the population in a range of areas, including digital skills. A key platform is the Korean Massive Open Online Courses; since its launch in 2015, over 1,700 accredited courses at the higher education level have been developed through partnerships with local universities, with a significant share of them focused on advanced digital courses such as machine learning, AI navigation and perception, and mathematics for data scientists. Uzbekistan could work with local universities and technical training institutions to launch platforms for open online courses or distance learning degrees. The Ministry of Higher Education could lead efforts to develop and accredit online courses, particularly courses focused on 4IR-related skills and implement relevant programs over the next 12 months.

(v) **Promote the use of innovative technologies to strengthen training delivery.** The surveys conducted found that 56% of training institutions in Uzbekistan use online self-learning modules to deliver training and only a third use virtual reality or augmented reality mechanisms. The use of innovative education technologies could allow training institutions to strengthen training delivery. For instance, researchers at the University of New South Wales in Australia developed a virtual reality platform that allows construction workers to navigate life-threatening scenarios using a computer or virtual reality headset. Collaboration with the private sector could help to address any potential resource constraints faced by training institutions in Uzbekistan in the adoption of 4IR technologies. Such efforts could be led by the Ministry of Higher Education and implemented over the next 12 to 36 months.

(vi) **Adopt programs to strengthen industry knowledge of trainers.** The employer surveys revealed that employers in the construction and textile and garment manufacturing industries face challenges in identifying and hiring graduates that are sufficiently prepared for the job by their training. Consultations with government stakeholders further suggest a need for closer collaboration between industry and training institutions on the delivery of training, to ensure that graduates have the practical skills required to take on jobs. To ensure that trainers have industry-specific knowledge and improve the quality of training, Uzbekistan could take reference from Malaysia's Faculty Industry Attachment program under which lecturers undergo an industrial attachment in related organizations to gain an in-depth understanding of the current demands of the industry. Similar efforts or programs could be led by the Ministry of Higher Education and implemented over the next 12 to 36 months.

(vii) **Develop targeted programs to ensure that women can benefit from 4IR.** Targeted programs are needed to overcome sociocultural norms in Uzbekistan and ensure that female workers can benefit from higher-skilled, better-paying technical jobs created by 4IR in the long term. Currently, while 45% of male tertiary students in Uzbekistan pursue science, technology, engineering, and mathematics (STEM) fields, a similar percentage of female workers are enrolled in education-related disciplines instead. In Pakistan, the ICTs for Girls program provides opportunities for girls and women to improve their digital literacy and employability. As part of this program, tens of thousands of girls and women from disadvantaged segments of society are provided digital infrastructure with state-of-the-art machines in fully broadband supported environments. Policy makers in Uzbekistan could consider similar initiatives to strengthen the capabilities and interest of women in collaboration with international partners. Such efforts could be led by the Ministry for Innovation Development and implemented over the next 12 to 36 months.

(viii) **Develop skilling and labor support programs for digital freelancers.** One key aspect of the digital economy and 4IR is the proliferation of digital freelancers. There is significant potential for workers in Uzbekistan to find quality jobs as digital freelancers with the rise of the global freelance economy against the backdrop of the COVID-19 pandemic. Various initiatives could be pursued to strengthen the digital freelancing capabilities of the workforce and the support infrastructure for freelancers. In particular, the

Ministry of Employment and Labor Relations and Ministry for Innovative Development could lead efforts to create an enabling environment for digital freelancers through fiscal incentives and micro-certification programs. This could be implemented over the next 12 to 36 months.

While these recommendations apply to both the textile and garment manufacturing and construction industries, a set of priorities unique to each industry should be considered when implementing the respective recommendations.

Textile and Garment Manufacturing Industry

Around 71% of textile and garment manufacturing firms surveyed indicated a good understanding of 4IR technologies. However, it appears that training institutions face challenges in offering sufficient 4IR-related courses. For instance, while 55% of textile and garment manufacturing firms have adopted autonomous robots, only 10% of training institutions offer relevant courses. This challenge is likely to be exacerbated as 4IR changes the task profiles of jobs in the industry and skills needs. Creative thinking and/or design and digital and/or ICT skills will become the most valued skills by employers by 2025, with 4IR adoption. The development of sectoral 4IR adoption plans to coordinate technology adoption and skills development is therefore particularly critical for the textile and garment manufacturing industry to ensure that the curricula and courses provided by training institutions are aligned with the rapidly changing skills needs of employers. At the same time, programs to strengthen the industry knowledge of trainers would also be necessary to ensure that trainers are equipped to deliver courses in these new skills areas and technologies. These efforts could be complemented by programs to strengthen the training capabilities of employers.

Construction Industry

Firms in the construction industry vary widely in their understanding of 4IR with 35% reporting an advanced understanding. Interviews with in-market experts suggest that usage of 4IR technologies in Uzbekistan's construction industry is limited to larger firms involved in large-scale infrastructure projects, while smaller firms (e.g., subcontractors) see less need for such technologies and lack the resources to deploy them. Programs to build a strong awareness of digital tools available will therefore be particularly critical for small firms in the construction industry. In addition, programs to expand the use of AR/VR technologies in training delivery and increase the share of women in technical roles could be particularly useful for Uzbekistan's construction industry. Virtual reality-based training is particularly relevant for industries such as construction, in which untrained workers face a high risk of accidents but only 33% of training institutions use VR/AR technologies to deliver training in 2020 (based on survey conducted in 2021).

1 | The Industry 4.0 Skills Challenge

This chapter investigates the demand and supply of skills driven by the adoption of Fourth Industrial Revolution (4IR) or Industry 4.0 technologies for both the textile and garment manufacturing and construction industries in Uzbekistan. The analysis uses a range of data, including employer surveys, expert interviews, online job board data, and national labor market statistics; and projects implications up to 2025.

The analysis reveals the progress that Uzbekistan has made in building a good awareness of 4IR technologies and their applications among enterprises and highlights the areas in which further progress can be made. It demonstrates that the adoption of 4IR technologies will change the types of jobs available in these industries and the skill needs of employers significantly in the long-term. The analysis highlights the importance of aligning industry development policies with skills development strategies in enabling Uzbekistan's transition to 4IR.

The research points to the potential of 4IR to create net job gains in both industries. Close to 100,000 more jobs (29% of the 2020 textile and garment manufacturing industry workforce) over and beyond business-as-usual growth rates, are expected to be created in the textile and garment manufacturing industry by 2025; and over 334,000 more jobs (25% of the 2020 construction industry workforce) are expected to be created in the construction industry by 2025, if firms fully adopt 4IR technologies by then. While manual jobs will continue to form the largest proportion of jobs in both industries, the tasks undertaken by the employee and skills required will change. The percentage of time spent on physical, routine tasks will change and employees will be required to undertake more analytical tasks. Digital and/ or information and communication technology skills, as well as creative thinking and/or design skills, will therefore be increasingly valued by employers. One important finding of this research is that graduates hired by firms in the textile and garment manufacturing and construction industries at present are not adequately prepared for their roles so that firms need to invest substantially in worker training. Targeted measures will be needed to address this to ensure that Uzbekistan's firms and workers can reap the gains of Industry 4.0.

A. Industry 4.0 and Its Relevance for Uzbekistan

The Fourth Industrial Revolution (4IR) or Industry 4.0 is poised to fundamentally change the future of work. 4IR can be described as the advent of "cyber-physical systems" involving entirely new capabilities for people and machines,[1] wherein new technologies, such as the Internet of Things (IOT), artificial intelligence (AI), additive manufacturing, robotics, and Big Data analysis among others, become embedded within societies. 4IR is fundamentally different from past industrial revolutions in its potential implications for economies and the workforce.

---

[1] World Economic Forum. What is the Fourth Industrial Revolution? https://www.weforum.org/agenda/2016/01/what-is-the-fourth-industrial-revolution/.

What could 4IR mean for Uzbekistan? The Government of Uzbekistan recognizes the potential of 4IR and has adopted a series of strategies that set out Uzbekistan's vision of building a vibrant knowledge-based economy supported by a skilled workforce with strong digital literacy. The Five-Area Development Strategy, 2017–2021 sets out priorities to improve the system of lifelong learning, increase access to quality education, improve workforce employability, and stimulate research and innovation (*The Tashkent Times* 2017). The Digital Uzbekistan 2030 development road map sets out plans to encourage technology adoption across all sectors and improve the digital literacy of the workforce (Government of Uzbekistan 2020a).

Understanding how the skills landscape is likely to change under 4IR is becoming more difficult in the face of the rapid pace at which technology is developing and being adopted. This is particularly so as these changes are accelerated against the backdrop of the coronavirus disease (COVID-19) pandemic. This means traditional approaches of assessing skill gaps, often relying on time intensive processes to collect data that quickly become outdated, may no longer be suitable. This study explores a new approach to understanding the labor market implications of 4IR. Some of the key design aspects are as follows:

(i)   **Use of primary and secondary local data.** This study utilizes a variety of local data sources, including data from the State Committee of the Republic of Uzbekistan on Statistics; surveys conducted on employers in the textile and garment manufacturing and construction industries as well as training institutions in Uzbekistan; and interviews with local experts and key stakeholders. A summary of the primary data sources used is in Table 1.

Table 1: Primary Data Sources Used

Employer surveys	51 textile and garment manufacturing firms and 51 construction firms in Uzbekistan were surveyed. The surveys were carried out by M-Vector.
Training institution surveys	A survey of 70 training institutions was undertaken in Uzbekistan. These include technical colleges and specialized secondary education providers that provide technical and vocation education and training (TVET) training as well as institutions of higher learning. Of the TVET training institutions surveyed, 98% train at least 100 students per year. The surveys were carried out by M-Vector.
Online job portal analysis	The analysis covered 52 job listings in the textile and garment manufacturing and 99 job listings in the construction industry scraped from the job portal "HeadHunter Uzbekistan" in June 2021.

Source: Asian Development Bank and AlphaBeta.

(ii)   **Use of current market information.** Given the rapid changes in the labor market, labor market surveys can become quickly obsolete. To provide an updated snapshot of skills needs, this study uses information on skill profiles for current jobs advertised in major online job portals.[2] Machine learning algorithms were applied to analyze data scrapped from local job portals to understand the skills demanded by employers in the two industries.

(iii)   **Focus on both demand and supply.** The study considers changes in the demand of skills brought about by the adoption of 4IR technologies, as well as the supply of skills, and the readiness of the training landscape to upskill and reskill workers for 4IR.

---

[2]   The analysis covered 52 job listings in the textile and garment manufacturing and 99 job listings in the construction industry scraped from the job portal "HeadHunter Uzbekistan" in June 2021.

B.   Industry Selection

Two industries were selected to conduct further analysis of the implications of 4IR adoption for the demand and supply of skills. A two-step methodology was used to select the industries.

(i)   **Shortlisting industries prioritized by the Government of Uzbekistan for future growth or for 4IR application.** This included reviewing the Five-Area Development Strategy, 2017–2021 for Uzbekistan, the Digital Uzbekistan 2030 road map, and the Uzbekistan 2035 development strategy, among other policy documents.

(ii)   **Scoring and ranking shortlisted industries.** These were done according to a set of criteria:

    (a)   How significant is the industry's contribution to the country's employment?

    (b)   Does it exhibit strong recent employment growth?

    (c)   Are its exports internationally competitive?

    (d)   How relevant are 4IR technologies to the industry?

    (e)   Is the relevant data available for the industry analysis?

The shortlisted industries were then tested with stakeholders across government, industry, and civil society during a consultation workshop conducted in May 2021. Based on this process, the following industries were selected for the analysis in Uzbekistan:

(i)   **Textile and garment manufacturing.** The textile and garment manufacturing industry has been identified in national growth plans such as the Five-Area Development Strategy, 2017–2021 (*The Tashkent Times* 2017). As an industry with huge untapped potential to drive long-term economic growth in Uzbekistan, the industry is currently at a nascent stage of development, employing only 1.9% of the country's workforce in 2018. There is opportunity for more jobs to be created in line with government plans, and the Government of Uzbekistan has set ambitious growth targets for the industry. The government aims to fully phase out exports of raw cotton and increase exports of textile products to $7 billion by 2025, up from $1.9 billion in 2020. The country's recent admission to the special incentive arrangement for sustainable development and good governance (GSP+) under the unilateral Generalized Scheme of Preferences (GSP) scheme by the European Union (EU) will increase the competitiveness of Uzbekistan's textile products in Europe and provide further growth impetus for the industry (Xin Hua 2021). Past research demonstrates the significant potential for 4IR technologies to be adopted in the textile and garment manufacturing industry, with productivity gains of up to 46% (McKinsey & Company 2018). As Uzbekistan embarks on various reforms and measures to increase the productivity and competitiveness of its textile and garment manufacturing industry, 4IR technologies such as IOT, AI, additive manufacturing, autonomous robots, and big data analytics can be leveraged to meet these growth objectives.

(ii)   **Construction.** The construction industry is one of the largest employers in Uzbekistan, contributing 9.2% to national employment in 2018. The industry has benefited from Uzbekistan's drive to improve transport, energy, and urban infrastructure as well as the high demand for housing (OECD 2019). The construction value chain in Uzbekistan is well-established and comprises engineering and design services, builders and contractors, and construction equipment, and material factories. Many small and medium-sized enterprises (SMEs) are active in the industry and the total number of enterprises in the industry increased by 85% between 2017 and 2021. These SMEs carried out 74% of the total volume of construction work in January 2021 (UZ Daily 2021a). The construction industry features strongly in national growth plans with the Strategy of Modernization and Innovative Development of the Construction Industry, 2021–2025 setting out plans for the adoption of innovative technologies and human capital development (Government of Uzbekistan 2020b). In particular, the adoption of 4IR technologies could bring significant benefit to the construction industry, with past research estimating that full-scale digitization could generate an estimated 12%–20% in annual cost savings (Buehler, Buffet, and Castagnino 2018). It is therefore timely for policy makers to consider how 4IR technologies can be integrated into the industry's growth strategies.

C. Textile and Garment Manufacturing

Industry 4.0 technologies pose significant potential for textile and garment manufacturers to increase their labor productivity. There are estimates that the full adoption of 4IR technologies across the textile and garment manufacturing value chain could reduce labor time by 40%–70% (McKinsey & Company 2018). The projections in Uzbekistan are consistent with these expectations. Textile and garment manufacturers in Uzbekistan have a strong understanding of 4IR technologies with 53% reporting an advanced understanding of 4IR technologies and their applications. Based on the survey, firms expect labor productivity (as measured by output per worker) to increase by 68% on average by 2025, with the adoption of 4IR technologies.

If these productivity gains are realized across the entire textile and garment manufacturing industry in Uzbekistan, a significant number of new jobs will be created by 2025. This report estimates that 100,000 net new jobs or the equivalent of 29% of the 2020 textile and garment manufacturing workforce could be created by 2025, over and beyond the number of jobs created by business-as-usual (BAU) growth. The 100,000 net new jobs created are a function of 247,000 jobs created and 147,000 jobs displaced with the adoption of 4IR technologies. In other words, this does not mean that no workers will be displaced by automation but that the number of jobs created by the overall growth of the industry due to 4IR adoption will exceed those displaced. However, the skills required in these new job roles created could be different from the skills required by textile and garment manufacturing workers at present. Creativity and design skills as well as digital and/or ICT skills will be increasingly valued by employers according to the employer surveys conducted. The development of strong human capital that embodies these skill sets will be in turn critical to ensure that the industry can adopt new 4IR technologies.

Relevance of Industry 4.0 to the Textile and Garment Industry

There are various 4IR technologies relevant to the textile and garment manufacturing industry, ranging from robotics technology to additive manufacturing processes that enable the mass customization of products.

Some key 4IR technologies and their applications in the textile and garment manufacturing industry include the following:

(i)   **Internet of Things.** The IOT refers to networks of sensors and actuators embedded in machines and other physical objects that connect with one another and the internet. The textile manufacturing process has many stages such as spinning, weaving, dyeing, printing, finishing, and fabric manufacturing, all of which require close monitoring. An IOT-integrated system, enabled by sensors and drones, can help to provide real-time data across these stages and identify potential bottlenecks. One example is an IOT-enabled weaving unit that can synchronize every stage of the weaving process from yarn processing and inventory to production monitoring, right up to shipment of the finished fabric. It can manage the yarn inventory and optimize the production schedule, among other functions.[3] Egyptian startup Garment IO uses IOT technologies to transform clothing factories. Every worker is given a smart terminal and an electronic card. Once a batch of work has been completed, such as sewing an order of shirts, the workers scan their card on the terminal and the tag attached to the bundle of shirts. The smart terminal logs what order the worker completed, how long it took them, and how many more orders they have left to complete. This information is then uploaded to a cloud that managers can access in real-time along with detailed breakdowns of how the factory line is performing. This allows factories to identify and eliminate any potential bottlenecks quickly (Waya 2019).

(ii) **Artificial intelligence.** Artificial intelligence gives machines the ability to learn and act intelligently and carry out a wide range of human-like processes. This means they can make decisions, carry out tasks, and even predict future outcomes based on what they learn from the data. For instance, AI technology can be used to carry out fabric grading, and quality checks on fabrics and ensure that the colors of the finished textile match with the originally designed colors. Researchers in Hong Kong, China developed the WiseEye, an AI-based automated textile inspection system to help manufacturers instantly detect defects in fabrics during production. WiseEye can reduce loss and wastage due to faulty textiles by 90% as compared to human inspection (Del Buono 2018). In Pakistan, firms such as Masood Textile Mills and Amami Clothing use CLO, a 3D garment simulation software, in their product development process. The software can be used to create virtual prototypes and modeling to replace physical prototypes and reduce the overall time and cost of the design process.[4] Retailers also use AI technologies to improve the online shopping experience. The AI-based startup Lalaland creates fully AI-generated fashion models that online shoppers can customize to look like themselves. By tailoring the "virtual fitting" experience, Lalaland's technology claims to be able to achieve a 15% increase in sales and a 10% reduction in returns for online retailers (Nicholls-Lee 2021).

(iii) **Additive manufacturing.** Additive manufacturing technologies or 3D printing produce physical objects from digital models by adding thin layers of material in succession. The process cuts down material waste and improves production efficiency. Additive manufacturing can be used to create customized clothing, complex designs in garments and accessories, and prototypes. For example, sportswear brand Adidas is leveraging 3D printing technology and robotics to produce footwear modelled to the exact contours of an individual runner's foot (Hanaphy 2020). Apparel brand Ministry of Supply can produce a customized blazer in just 90 minutes using 3D printing technology, while reducing fabric waste in production by about 35% compared to traditional techniques (*CB Insights 2022*).

(iv) **Autonomous robots.** Autonomous robots are intelligent machines capable of performing tasks with a high degree of autonomy. There are several applications of robotics in this sector, including using robotic arms for repetitive processes such as weaving and sewing, needle positioning, fabric adjustment. Sewing is currently the most labor-intensive step in creating a garment, accounting for more than half the total labor time per garment. The potential for labor reduction varies by garment type, but as much as 90% of the sewing processes can potentially be automated (ACT/EMP and International Labour Organization [ILO] 2017). Studies show that full automation can reduce the production time of a shirt by 18–33 times compared to manual sewing (Institute for Workers and Trade Unions 2020). Autonomous robots also help to streamline other processes in the textile manufacturing production chain. For instance, Lahore-based garment manufacturer, Combined Fabrics, uses advanced cutting equipment from Tukatech, a leading provider of fashion software and machinery (Tukatech 2021). The automated precision cutting machines allow employees to be deployed to other tasks and reduces wastage by 12%–14% per garment.

---

[3] Clarion Technology. How IOT Transforms the Way to a More Sustainable Textile Manufacturing. https://www.clariontech.com/blog/how-iot-transforms-the-way-to-a-more-sustainable-textile-manufacturing.

[4] Sources: Clo3d. Design Smarter. https://www.clo3d.com/; local expert.

(v) **Big Data analytics.** Big data analytics is the use of advanced analytic techniques on large, diverse data sets. In the textile and garment manufacturing industry, applications include predictive analytics for maintenance and repair of production lines, and analysis of consumer data to predict consumer buying patterns and fashion trends. Fashion retailers can use data analytics to improve the efficiency of their e-commerce business. Fashion tech company Virtusize enables online shoppers to buy the right size, either by measuring the clothes in their closet or by comparing specific brands and styles to their own (Virtusize 2021). Virtusize can increase average order values by 20% and decrease return rates by 30% by reducing uncertainty around size and fit. Similarly, online styling service Stitch Fix uses big data analytics to deliver personalized style recommendations to customers. All of Stich Fix's revenue results directly from its recommendations that combine data and machine learning with expert human judgment (Lake 2018).

For the employer surveys in Uzbekistan, 51 textile and garment manufacturing firms were surveyed. The employer survey shows that a significant proportion of textile and garment manufacturers have a strong understanding of 4IR technologies, with 53% indicating an advanced level of understanding of 4IR technologies and the potential benefits that such technologies could bring to their industry (Figure 1). Of the 51 firms, 55% indicated that firms in their supply chain, such as material suppliers and retailers, already have an advanced understanding of 4IR technologies and use these technologies in their operations (Figure 2). The strong awareness of 4IR technologies

Figure 1: Employers' Understanding of Industry 4.0 Technologies in the Textile and Garment Manufacturing Industry in Uzbekistan

More than 50% of firms in the textile and garment manufacturing industry believe that they have an advanced understanding of 4IR technologies

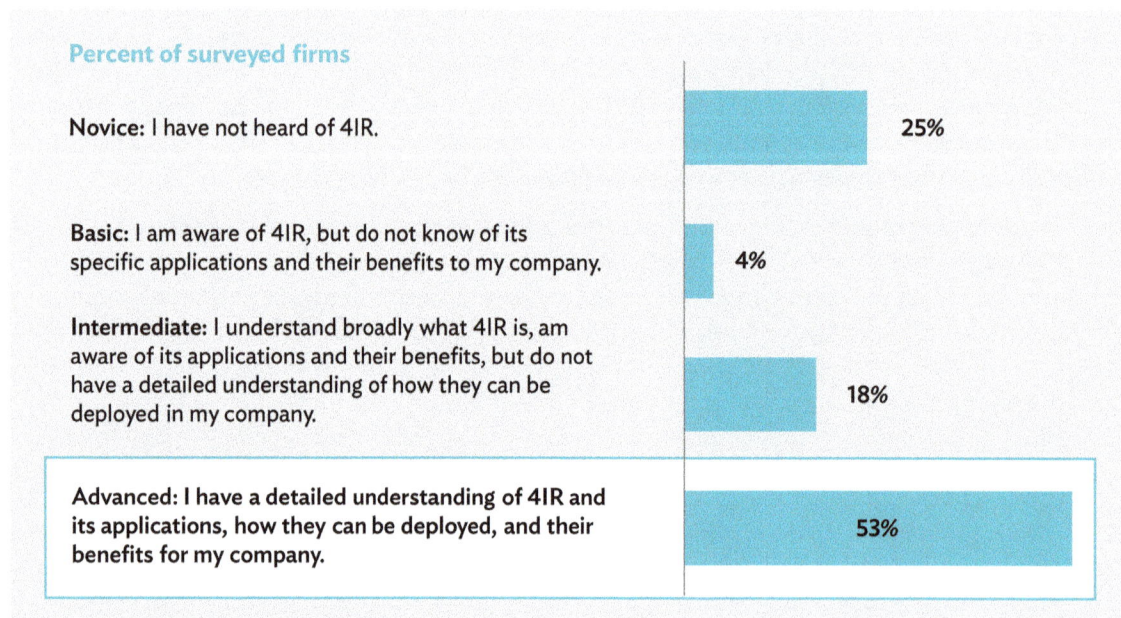

Percent of surveyed firms

Novice: I have not heard of 4IR. — 25%

Basic: I am aware of 4IR, but do not know of its specific applications and their benefits to my company. — 4%

Intermediate: I understand broadly what 4IR is, am aware of its applications and their benefits, but do not have a detailed understanding of how they can be deployed in my company. — 18%

Advanced: I have a detailed understanding of 4IR and its applications, how they can be deployed, and their benefits for my company. — 53%

4IR = Fourth Industrial Revolution.
Note: Based on survey of employers in the textile and garment manufacturing industry between June and September 2021 (n=51).
Source: Asian Development Bank (Sustainable Development and Climate Change Department).

Figure 2: Employers' Understanding of Industry 4.0 Technologies among Firms in the Textile and Garment Manufacturing Supply Chain in Uzbekistan

Fifty-five percent of textile and garment manufacturing firms feel that companies in their supply chain have an advanced understanding of 4IR technologies

Percent of surveyed firms

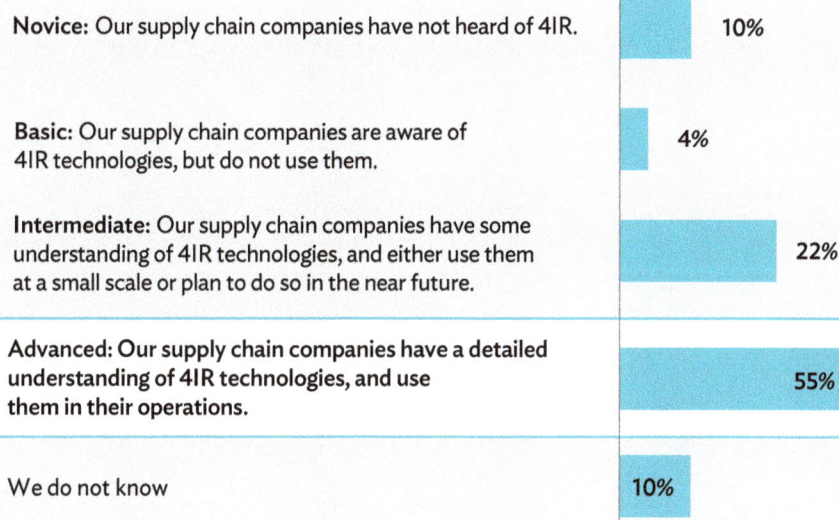

Novice: Our supply chain companies have not heard of 4IR.	10%
Basic: Our supply chain companies are aware of 4IR technologies, but do not use them.	4%
Intermediate: Our supply chain companies have some understanding of 4IR technologies, and either use them at a small scale or plan to do so in the near future.	22%
Advanced: Our supply chain companies have a detailed understanding of 4IR technologies, and use them in their operations.	55%
We do not know	10%

4IR = Fourth Industrial Revolution.
Note: Based on survey of employers in the textile and garment manufacturing industry between June and September 2021 (n=51).
Source: Asian Development Bank (Sustainable Development and Climate Change Department).

is backed by government policies to encourage Uzbekistan's textile and garment manufacturers to adopt innovative production methods. In particular, the 2017 Presidential Decree on Measures for the Accelerated Development of the Textile, Sewing, and Knitting Industry (UZ Daily 2017) sets out measures to modernize production processes; introduce advanced information and communication technology (ICT) technologies; as well as implement international standards to improve product quality in the textile and garment manufacturing industry. However, more than a quarter of firms surveyed have a limited understanding of 4IR technologies and their applications. These are likely to be SMEs that make up a substantial part of Uzbekistan's economy, employing close to 80% of the workforce (OECD 2017). Targeted programs would therefore need to be adopted to ensure gains from 4IR are equitably distributed.

Corresponding to their strong understanding of 4IR technologies and their applications, textile and garment manufacturing firms in Uzbekistan expect 4IR technologies to bring significant productivity gains. Past research estimates that through adopting relevant technologies in the textile and garment manufacturing industry, productivity could be increased by 21%–46% (McKinsey & Company 2018). In Uzbekistan's textile and garment manufacturing industry, firms are more optimistic and expect labor productivity to increase by 68% on average by 2025, with the adoption of 4IR technologies (Figure 3).

Figure 3: Expected Increase in Output Per Worker Due to Industry 4.0 Technologies in the Textile and Garment Manufacturing Industry in Uzbekistan, 2020–2025

The majority of firms expect labor productivity to increase by 50% to 100% in 5 years' time with the adoption of 4IR technologies

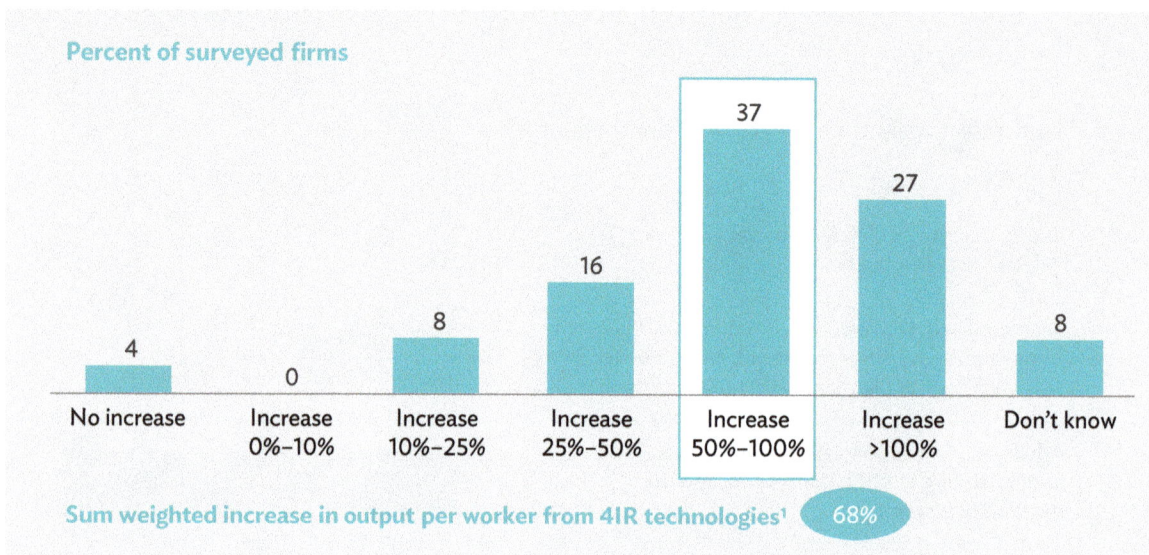

Percent of surveyed firms

No increase	Increase 0%–10%	Increase 10%–25%	Increase 25%–50%	Increase 50%–100%	Increase >100%	Don't know
4	0	8	16	37	27	8

Sum weighted increase in output per worker from 4IR technologies[1]   68%

4IR = Fourth Industrial Revolution.

Notes: Based on survey of employers in the textile and garment manufacturing industry between June and September 2021 (n=51). Calculated using sum-weighted average of output increase by the number of firms indicating different levels of expected increase in output, i.e., 0%, 0%–10%, 10%–25%, 25%–50%, 50%–100%, and over 100%. The midpoint of the range for each option for expected increase in output is used; for expected output increase of over 100%, the lower bound of 100% is used.

Source: Asian Development Bank (Sustainable Development and Climate Change Department).

The strong expectations of labor productivity gains are consistent with the plans of firms to significantly increase their adoption of 4IR technologies over the next 5 years. (Figure 4). Autonomous robots show the highest levels of adoption among surveyed firms in 2020; and the adoption of IOT technologies and big data analytics is expected to rise significantly by 2025. Eighty percent of surveyed firms in Uzbekistan currently use autonomous robots to some extent. For instance, UZTEX Group, one of the country's largest textile firms, uses yarn and fabric dyeing processes that are fully automated. The preparation and supply of dyes to boilers, including temperature and time modes, are fully automated and controlled by a central server. The high level of automation could be due to the relatively late development of Uzbekistan's textile and garment manufacturing industry and government policies in this area. In the early 2000s, many new modern facilities that specialized in manufacturing finished textile and garment products were built, driven by strong policies to attract foreign direct investments (Tursunov 2017). Since then, the government has also embarked on programs to modernize and re-equip the industry with up-to-date equipment and technologies (Tursunov 2017).

From 2020 to 2025, 67% of firms expect to deploy IOT technologies across all possible functions and 65% expect to use big data analytics across all possible functions. IOT technologies can be used to improve the efficiency and quality control of production processes. For instance, in the spinning process, which includes various intermediate processes, IOT integration can support the collection of data over the intermediate processes in

Figure 4: Current and Future Adoption of Relevant Industry 4.0 Technologies in the Textile and Garment Manufacturing Industry in Uzbekistan

Adoption of 4IR technologies will increase in the textile and garment manufacturing industry in the next 5 years

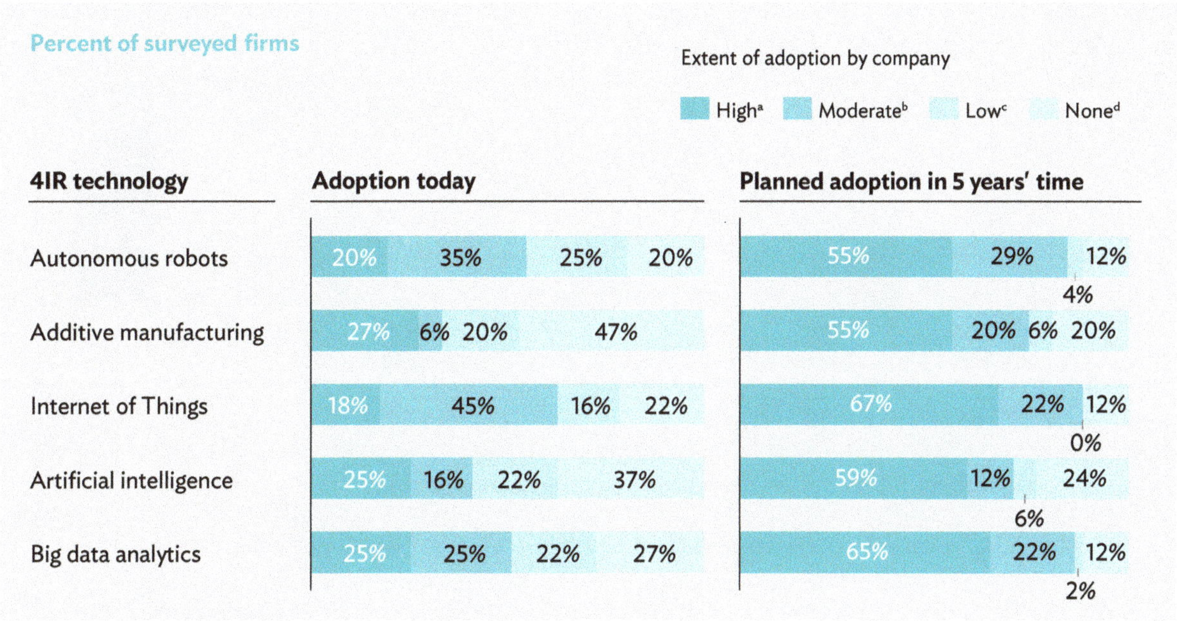

Percent of surveyed firms

Extent of adoption by company

■ High[a]   ■ Moderate[b]   ■ Low[c]   None[d]

4IR technology	Adoption today				Planned adoption in 5 years' time			
	High	Moderate	Low	None	High	Moderate	Low	None
Autonomous robots	20%	35%	25%	20%	55%	29%	12%	4%
Additive manufacturing	27%	6%	20%	47%	55%	20%	6%	20%
Internet of Things	18%	45%	16%	22%	67%	22%	12%	0%
Artificial intelligence	25%	16%	22%	37%	59%	12%	24%	6%
Big data analytics	25%	25%	22%	27%	65%	22%	12%	2%

4IR = Fourth Industrial Revolution.

[a] "High": Firm has fully deployed the technology across all possible functions in the enterprise and/or has plans to fully deploy the technology across all possible functions in the future.

[b] "Moderate": Firm has implemented the technology, but not fully deployed across all possible functions in the enterprise and/or plans to implement the technology across a few functions in the future.

[c] "Low": Firm is experimenting with the technology at a limited scale within the enterprise and/or plans to experiment with the technology in the future.

[d] "None": Firm has not used technology at all within the enterprise and/or has no plans to use the technology in the future.

Note: Based on survey of employers in the textile and garment manufacturing industry between June and September 2021 (n=51).

Source: Asian Development Bank (Sustainable Development and Climate Change Department).

real-time to control production and quality as well as identify potential bottlenecks. Big data analytics are used by textile and garment manufacturing firms to analyze consumer data to predict consumer buying patterns and fashion trends, among other functions, and will become increasingly critical as consumer buying patterns change rapidly with the growth of e-commerce.

Based on a global survey of executives, the COVID-19 pandemic has led to firms accelerating the digitization of their customer and supply-chain interactions and their internal operations by 3–4 years (McKinsey and Company 2020). However, only 50% of textile and garment manufacturing firms surveyed in Uzbekistan assess that COVID-19 will accelerate the adoption of 4IR technologies in their industry (Figure 5). Among firms that expect the adoption of 4IR technologies to be accelerated by COVID-19, a strategic shift made by management to move toward greater digitization is seen as the main reason. Meanwhile, 24% of firms disagreed that COVID-19 will accelerate the adoption of 4IR technologies. The mixed reaction among firms could be due to the adverse

Figure 5: Perceptions on Impact of the COVID-19 Pandemic on Adoption of Industry 4.0 Technologies in the Textile and Garment Manufacturing Industry in Uzbekistan

Around half of employers believe that the COVID-19 pandemic has accelerated or will accelerate the use of 4IR technologies

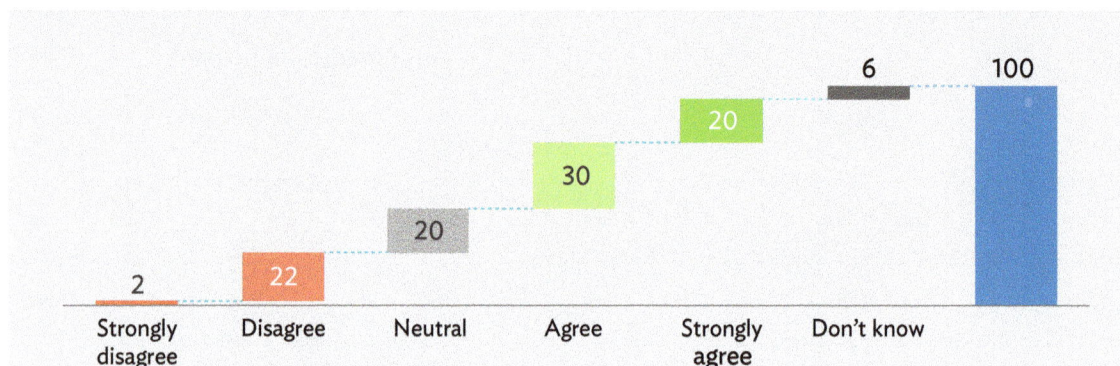

Common reasons for accelerated adoption

Strategic shift towards greater digitization by company's management.
Lack of labor due to movement restrictions necessitates more automation and shifting of activities to digital means.

Note: Based on survey of employers in the textile and garment manufacturing industry between June and September 2021 (n=51).
Source: Asian Development Bank (Sustainable Development and Climate Change Department).

impact of COVID-19 on textile and garment manufacturers around the world even as Uzbekistan's textile industry has grown. In Uzbekistan, the industry grew during the pandemic as new products such as protective masks and clothing were manufactured and exported to nearby countries (Fibre2Fashion 2020). In other textile manufacturing economies such as Viet Nam however, shipment delays and factory closures forced by movement restrictions had significant impact on manufacturers (Fibre2Fashion 2021). This could make firms in Uzbekistan less willing to invest capital in new technologies against an uncertain global economic climate.

Skills Demand Analysis

Job Implications

To determine the impact that the adoption of 4IR technologies will have on employment in Uzbekistan's textile and garment manufacturing industry in 2025, the displacement and productivity effects of adopting 4IR technologies were estimated to determine the net change in job numbers (Figure 6):

(i) **Displacement effect.** This refers to the number of jobs that could potentially be lost due to automation using 4IR technologies. 147,000 jobs or 43% of the current workforce size in Uzbekistan's textile and garment manufacturing industry could potentially be displaced due to the adoption of 4IR technologies in 2025.

Figure 6: Estimated Impact of Industry 4.0 on Number of Jobs in the Textile and Garment Manufacturing Industry in Uzbekistan, 2020–2025

The adoption of 4IR technologies could lead 29% more jobs in 5 years, time in the textile and garment manufacturing sector

Percent of jobs impacted due to displacement and productivity effects of 4IR in 5 years' time

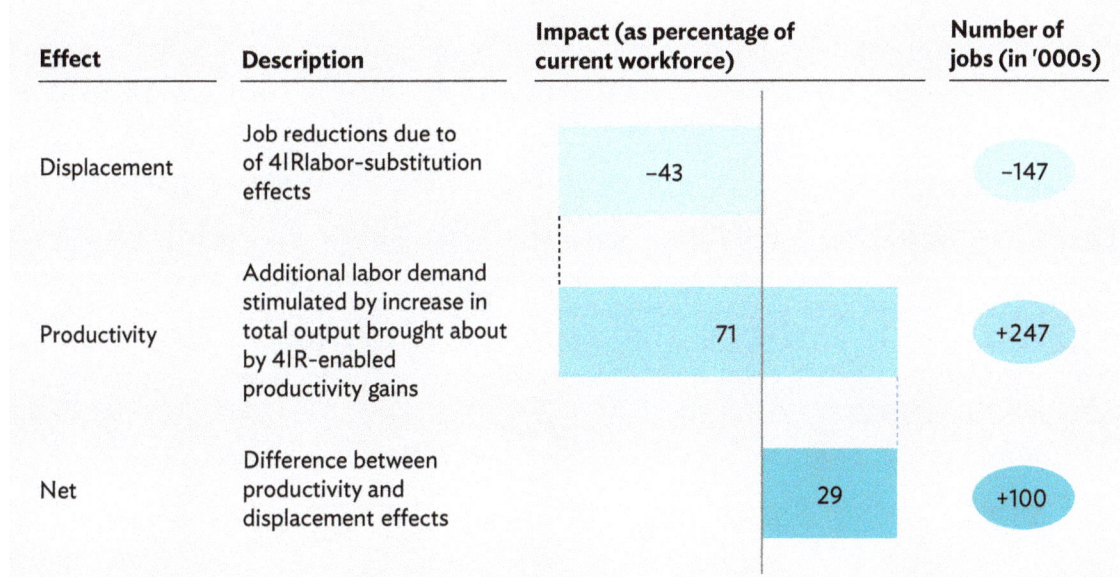

Effect	Description	Impact (as percentage of current workforce)	Number of jobs (in '000s)
Displacement	Job reductions due to of 4IRlabor-substitution effects	–43	–147
Productivity	Additional labor demand stimulated by increase in total output brought about by 4IR-enabled productivity gains	71	+247
Net	Difference between productivity and displacement effects	29	+100

4IR = Fourth Industrial Revolution.
Notes: Based on survey of employers in the textile and garment manufacturing industry between June and September 2021 (n=51). Industry employment data is from State Committee of the Republic of Uzbekistan on Statistics. Labor Market. https://stat.uz/en/official-statistics/labor-market and output data is from State Committee of the Republic of Uzbekistan on Statistics. Industry. https://stat.uz/en/official-statistics/industry.
Source: Asian Development Bank (Sustainable Development and Climate Change Department).

(ii) **Productivity effect.** This refers to the job gains due to improved productivity from technology adoption, which increases the potential total output of the industry and the demand for labor. If policies that encourage full 4IR adoption are implemented, up to 247,000 jobs could be created in Uzbekistan's textile and garment manufacturing industry by 2025 due to the productivity effect, based on our modeling.

This report estimates that the jobs created by the productivity effect will exceed those displaced, creating close to 100,000 new jobs or the equivalent of 29% of the 2020 textile and garment manufacturing workforce in net job gains in 2025, over and beyond workforce changes due to BAU growth. There are however two caveats to realizing these 4IR job gains. First, these gains assume that firms in Uzbekistan fully adopt 4IR technologies. However, more than a quarter of firms in the textile and garment manufacturing industry have a limited understanding of such technologies and their applications as of 2020. Second, new jobs created might not be identical to the jobs displaced and could require different skill sets. As such, those who risk losing their jobs may not seamlessly move into new jobs being created without some level of reskilling.

In interpreting the gains from 4IR adoption, it is important to understand that the 100,000 net jobs estimated will be created over and beyond new jobs due to BAU growth. In other words, the textile and garment manufacturing workforce grew at approximately 1% per annum[5] from 2015 and 2020; and if this growth were extrapolated up to 2025, an additional 17,000 jobs would have been created even without adopting 4IR technologies. However, the full adoption of 4IR technologies could see 117,000 more jobs in 2025 than in 2020.

Box 1: Estimating Employment Changes Due to Adoption of Industry 4.0 Technologies

To determine the impact that the adoption of Fourth Industrial Revolution (4IR) technologies will have on employment in Uzbekistan's textile and garment manufacturing industry from 2020 to 2025, data on the business-as-usual (BAU) growth of the industry and responses from the employers' survey were utilized. The displacement and productivity effects of adopting 4IR technologies were estimated to determine the net change in jobs.

- **Displacement effect.** This refers to the number of jobs potentially lost due to automation using 4IR technologies. It assumes that as output per worker increases due to the adoption of 4IR technologies, fewer workers would be needed to produce the same amount of output under a BAU scenario in 2025. The BAU amount of output in 2025 was calculated based on growth rates before the coronavirus disease (COVID-19) pandemic struck, i.e., from 2014 to 2019, while pre-COVID-19 industry labor productivity growth rates from 2014 to 2019 were used to calculate the BAU labor productivity (i.e., output per worker) in 2025. The expected labor productivity increase from the adoption of 4IR technologies, on top of the expected BAU labor productivity in 2025, was obtained from the employer survey to calculate the displacement effect.

- **Productivity effect.** This refers to the job gains due to improved productivity from technology adoption, which increases the potential total output of the industry. For instance, the improved ease in production or the manufacturing of higher quality goods can both lead to higher total industry output and corresponding job creation. It assumes that (i) the market can completely absorb the higher output produced; and (ii) firms can produce at the original cost notwithstanding the higher productivity (e.g., pay wages based on the BAU productivity of workers prior to 4IR adoption). Using the labor productivity increase estimated from the survey, on top of the labor productivity increase that would have taken place at BAU in the 2025, the industry's potential total output with 4IR adoption in 2025 was calculated, and the number of additional workers that could be hired if labor productivity had remained at BAU levels 2020 to 2025 was determined..

- **Net gains.** The net gains were determined by taking the net of the increase in jobs created by the productivity impact and decrease in jobs created by the displacement impact.

Source: Asian Development Bank (Sustainable Development and Climate Change Department).

To better understand the differences between the jobs displaced and the jobs created, it is important to recognize that each industry is characterized by various occupational roles carrying out different tasks, and the impact of 4IR technologies on these roles is uneven. For example, manual roles such as a sewing machine operator or weaver operating traditional hand-powered weaving machines typically face a higher risk of automation. This, while technology adoption could increase the number of technical roles as more workers are needed to operate and repair more advanced equipment or to use digital platforms. In this report, jobs in Uzbekistan's textile and garment manufacturing industry are split into five occupational groups (Table 2).

Across all occupational groups, the proportion of employers that expect the number of jobs to increase in 2025 exceed the proportion of employers that expect jobs to decrease (Figure 7). There could be three reasons for

---

[5]   Calculated using industry and workforce data from State Committee of the Republic of Uzbekistan on Statistics. As a sectoral workforce growth figure was not available, the growth of the "industry" workforce from 2015 to 2020 was used to estimate the sectoral workforce growth rate.

Table 2: Occupational Groups in the Textile and Garment Manufacturing Industry

Occupational Group	Possible Job Titles
1 **Technical**	• Sewing production line operator • Engineering technician • Garment technologist
2 **Managerial**	• Chief executive officer • Factory floor manager
3 **Customer-facing**	• Marketing manager • Retail sales associate
4 **Administrative**	• Secretary • Finance executive
5 **Elementary and/or manual jobs**	• Sewing machine operator • Weaver

Source: Asian Development Bank (Sustainable Development and Climate Change Department).

Figure 7: Employers' Expectations on Impact of Industry 4.0 on the Number of Jobs in the Textile and Garment Manufacturing Industry in Uzbekistan, 2020–2025

In the textile and garment manufacturing industry, the number of jobs in all occupational groups are expected to increase in 5 years' time

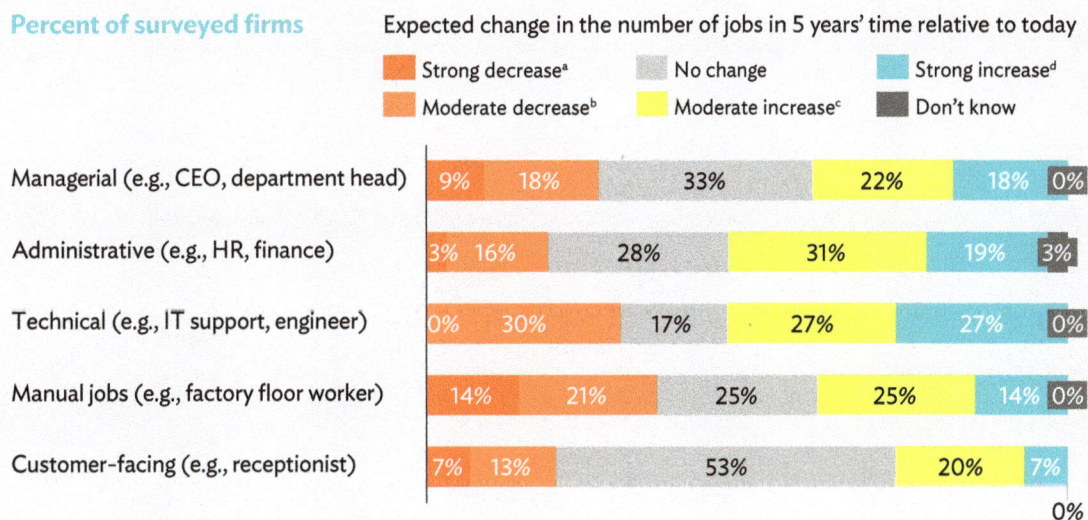

CEO = chief executive officer, HR = human resources, IT = information technology.
a   Greater than or equal to 50% decrease in number of jobs.
b   Less than 50% decrease in number of jobs.
c   Less than 50% increase in number of jobs.
d   Greater than or equal to 50% increase in number of jobs.
Note: Based on survey of employers in the textile and garment manufacturing industry between June and September 2021 (n=51).
Source: Asian Development Bank (Sustainable Development and Climate Change Department).

this. First, Uzbekistan's textile and garment manufacturing industry is growing rapidly, and employers could expect their workforce to increase overall across all occupational groups. Second, a large proportion of textile and garment manufacturers in Uzbekistan already use 4IR technologies and expect to continue to reap productivity gains without displacing workers. Third, expert interviews revealed that handcrafted products are associated with good quality and uniqueness in Uzbekistan. Some manufacturers see such products as their unique selling point and would therefore continue to use manual labor instead of machines for some functions.

Notwithstanding these factors, the composition of the textile and garment manufacturing workforce will change by 2025 (Figure 8). The proportion of manual jobs will decrease by over 2 percentage points, from 57.1% to 55% while the proportion of technical jobs will increase by 1 percentage point. If the assumption that employers currently expect to hire more workers across all categories as they expect the industry rapidly holds true, the shift will become larger once the industry's overall growth slows and the drop in the proportion of manual workers will become more significant. The ILO found that in economies with mature textile industries, textile and garment manufacturing workers are one of the most vulnerable to extensive technological displacement (ILO 2016). Another ILO assessment in the Association of Southeast Asian Nations (ASEAN) shows that 64%–88% of workers in this sector in ASEAN countries are at risk of being displaced by automation, the highest compared to other manufacturing sectors. This is likely due to the industry consisting of repetitive and mundane jobs that are replaceable by programmed machinery and engineering advancements. That said, such replacement could in fact lead to an increase in work satisfaction as machines take over a greater share of full routine tasks as the monotonous, automatable tasks performed by typically low-skilled workers have also been found to be the least satisfying tasks to perform (AlphaBeta 2017).

Figure 8: Composition of Jobs in 2020 and by 2025 in the Textile and Garment Manufacturing Industry in Uzbekistan

In the textile and garment manufacturing industry, the number of jobs in all occupational groups are expected to increase in 5 years' time

Weighted average percentage share of employees by occupational group in surveyed firms[a]

🔴 Negative shift
🟢 Positive shift

Occupational group	Share today	Share in 5 years' time[b]	Percentage shift
Manual jobs	57.1%	55.0%	−2.1%
Customer-facing	3.7%	3.6%	−0.1%
Managerial	9.5%	9.6%	+0.1%
Administrative	13.8%	14.8%	+1.0%
Technical	16.0%	17.1%	+1.1%

[a] Average share of employees in surveyed firms is weighted by the number of employees in each firm, as indicated by respondents. Percentages may not add up to 100% due to rounding.

[b] The change in the number of workers in each job type is based on the number of firms indicating different levels of changes in number of jobs, i.e., "strong increase," "moderate increase," "no change," "moderate decrease," "strong decrease." The midpoint of the range for each option for expected change is used. For expected increase or decrease of over 50%, the low bound of 50% was used.

Note: Based on survey of employers in the textile and garment manufacturing industry between June and September 2021 (n=51).

Source: Asian Development Bank (Sustainable Development and Climate Change Department).

Based on the current growth expectations, the net job gains created by the adoption of 4IR technologies are expected to be distributed across male and female workers. This is due to the relatively large share of manual workers in the industry overall, and high proportion of female workers in manual jobs (Figure 9). However, this might not reflect the full picture of the types of jobs that will be created for male and female workers. Currently, while 45% of male tertiary students in Uzbekistan pursue science, technology, engineering, and mathematics (STEM) fields, a similar percentage of female workers are enrolled in education-related disciplines (ILO 2021). This suggests that women are less likely to be qualified for the technical, more skilled roles created by 4IR even if they benefit from the overall expansion of the industry. Policy makers in Uzbekistan can consider initiatives to build stronger digital literacy among women, as well as encourage female participation in STEM to ensure that both male and female workers benefit from 4IR.

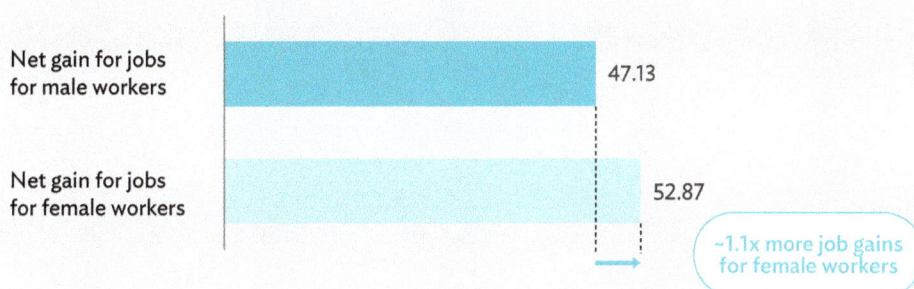

Figure 9: Estimated Net Job Gains by Gender from Industry 4.0 Adoption in the Textile and Garment Manufacturing Industry in Uzbekistan, 2020–2025

Job gains are expected to be distributed across male and female workers in textile and garment manufacturing industry

Estimated number of net jobs created by gender (in thousands)

Net gain for jobs for male workers — 47.13

Net gain for jobs for female workers — 52.87

~1.1x more job gains for female workers

4IR = Fourth Industrial Revolution, PBS = Pakistan Bureau of Statistics.
Note: Based on survey of employers in the textile and garment manufacturing industry between June and September 2021 (n=51). Industry employment data is from State Committee of the Republic of Uzbekistan on Statistics. Labor Market. https://stat.uz/en/official-statistics/labor-market.
Source: Asian Development Bank (Sustainable Development and Climate Change Department).

Task Implications

To understand the impact of the adoption of 4IR technologies on jobs and skills demand, it is important to understand that technology does not automate jobs, but rather individual tasks or combinations of tasks. This research examines five types of tasks linked to jobs in the textile and garment manufacturing industry and how they could be impacted by 4IR:

(i)   **Routine physical.** These tasks involve repetitive and predictable physical work. For example, a seamstress creating a handsewn traditional garment.

(ii) **Routine interpersonal.** These tasks involve predictable interactions with other people. For example, a retail sales assistant serving a customer.

(iii) **Nonroutine physical.** These tasks involve physical work that is not repetitive or predictable. For example, a mechanic diagnosing and repairing factory equipment.

(iv) **Nonroutine interpersonal.** These tasks involve complex or creative interactions with other people. For example, supervising others or making speeches or presentations.

(v) **Analytical.** These are tasks that vary significantly and involve a strong thinking and analytical component. They predominantly involve computers or other technological equipment.

Employers were asked to estimate how much time an average employee spent on each task type in an average work week in 2020 and predict how that would change in 5 years, or between 2020 and 2025, with the adoption of 4IR technologies. The analysis revealed that despite close to three quarters of firms having a good understanding of 4IR technologies, routine, physical tasks take up close to 70% of an average worker's work week in 2020. This is expected to drop significantly by 2025 as more firms adopt technologies such as autonomous robots across a range of functions (Figure 10). The amount of time spent on analytical tasks is expected to increase as more advanced equipment and software is deployed in the manufacturing process.

Figure 10: Time Spent by Employees on Tasks at Work in 2020 and by 2025 in the Textile and Garment Manufacturing Industry in Uzbekistan

Adoption of 4IR technologies is expected to shift the distribution of weekly working hours away from the routine physical tasks

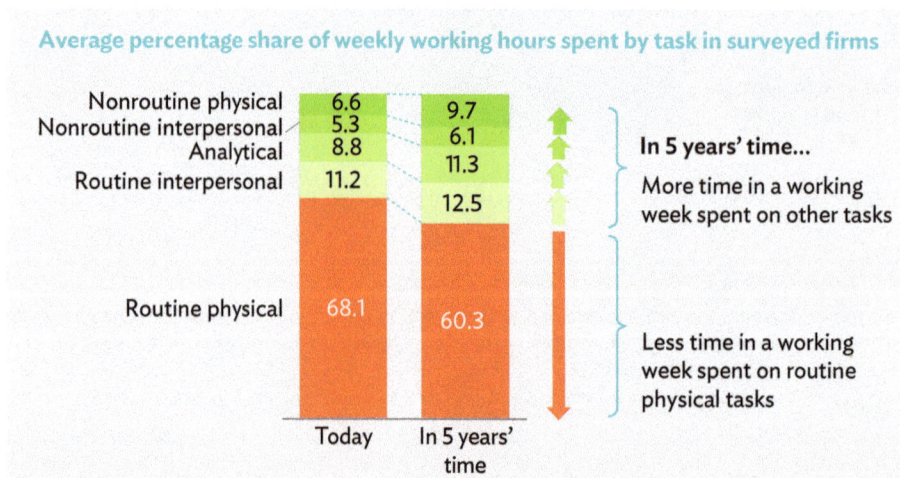

Average percentage share of weekly working hours spent by task in surveyed firms

	Today	In 5 years' time
Nonroutine physical	6.6	9.7
Nonroutine interpersonal	5.3	6.1
Analytical	8.8	11.3
Routine interpersonal	11.2	12.5
Routine physical	68.1	60.3

In 5 years' time...
More time in a working week spent on other tasks

Less time in a working week spent on routine physical tasks

Notes: Based on survey of employers in the textile and garment manufacturing industry between June and September 2021 (n=51). Figures include rounding adjustments.
Source: Asian Development Bank (Sustainable Development and Climate Change Department).

Skills Implications

The task shifts will impact the skills that are valued by employers in the textile and garment manufacturing industry. This analysis considers 10 categories of skills as set out in Table 3.

Table 3: Categories of Skills Considered in the Analysis

No.	Skill	Definition
1	**Creative thinking and/or design**	Ability to develop, design, or creating new applications, ideas, relationships, systems, or products
2	**Critical thinking**	Ability to use logic and reasoning to identify the strengths and weaknesses of alternative solutions, conclusions or approaches to problems
3	**Adaptive learning**	Ability to pick up new skills as demanded by the job
4	**Complex problem solving**	Ability to identify complex problems and review related information to develop and evaluate options and implement solutions
5	**Digital and/or ICT skills**	Ability to design, set-up, operate, and correct malfunctions involving application of machines or technological systems
6	**Numeracy**	Ability to add, subtract, multiply, or divide quickly and correctly and use mathematics to solve problems
7	**Written communication**	Ability to read and understand information and ideas presented in writing, and to communicate information and ideas in writing
8	**Verbal communication**	Ability to communicate information and ideas clearly by talking to other
9	**Management**	Ability to motivate, develop, and direct people as they work, and to identify the best people for the job
10	**Social and interpersonal**	Ability to work with people to achieve goals

ICT = information and communication technology.
Source: Asian Development Bank (Sustainable Development and Climate Change Department).

For 2020, employers ranked critical thinking as the most important skill followed by management and social and interpersonal skills. As the ranking was derived from an analysis of online job listings, it is likely that these are skewed toward employers in white collar roles and not indicative of the skills required of a manual worker. Nevertheless, the employer survey reveals that creative thinking and design skills and digital and/or ICT skills will become more important to employers for all categories of workers in 2025 (Figure 11). In the textile and garment manufacturing context, an increased demand for creative thinking and design skills could imply a greater demand for workers able to design new products or carry out digital marketing campaigns. Among the skills prioritized by employers in the next 5 years, digital and/or ICT skills are areas in which employers also see a strong need for reskilling (Figure 12). Conversely, while employers see workers to be lacking in numeracy skills in 2020, they find these skills to be less critical in 2025. Of the employers who feel that a step-up from basic proficiency in digital and/or ICT skills is needed, 91% would like to see a step-up to advanced proficiency and only 9% feel that step-up to intermediate proficiency would be sufficient.

Figure 11: Importance of Skills in 2020 and for Industry 4.0 Adoption by 2025 in the Textile and Garment Manufacturing Industry in Uzbekistan

There is a significant change in skills perceived as important by employers over the next 5 years

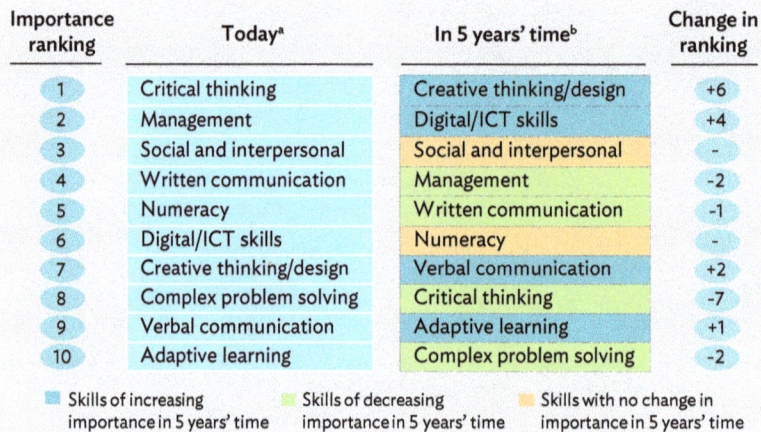

Importance ranking	Today[a]	In 5 years' time[b]	Change in ranking
1	Critical thinking	Creative thinking/design	+6
2	Management	Digital/ICT skills	+4
3	Social and interpersonal	Social and interpersonal	-
4	Written communication	Management	-2
5	Numeracy	Written communication	-1
6	Digital/ICT skills	Numeracy	-
7	Creative thinking/design	Verbal communication	+2
8	Complex problem solving	Critical thinking	-7
9	Verbal communication	Adaptive learning	+1
10	Adaptive learning	Complex problem solving	-2

■ Skills of increasing importance in 5 years' time   ■ Skills of decreasing importance in 5 years' time   ■ Skills with no change in importance in 5 years' time

4IR = Fourth Industrial Revolution, ICT = information and communication technology.

[a] Evaluated using employer survey and supported by job portal data.
[b] Evaluated using the employer survey.

Notes: Based on survey of employers in the textile and garment manufacturing industry (n=51); job data on the textile and garment manufacturing industry from the job portal HeadHunter Uzbekistan (accessed June 2021).

Source: Asian Development Bank (Sustainable Development and Climate Change Department).

Figure 12: Required Step-Up in Level of Proficiency of Employees' Skills from 2020 for Industry 4.0 Adoption by 2025 in the Textile and Garment Manufacturing Industry in Uzbekistan

To be 4IR-ready, workers would require significant proficiency leaps in critical thinking and numeracy skills

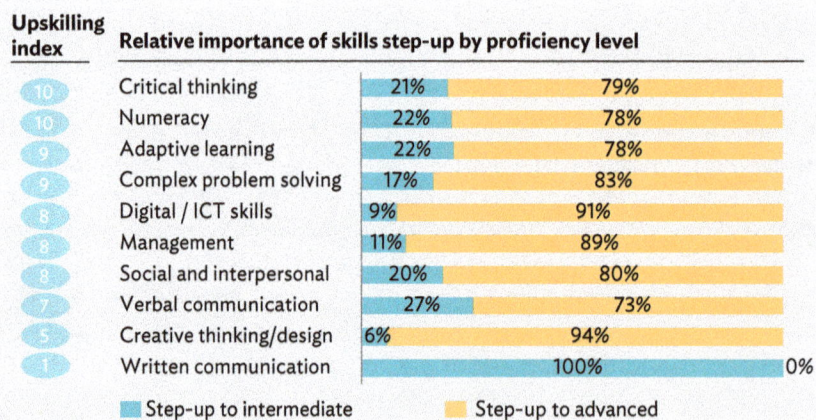

Upskilling index	Relative importance of skills step-up by proficiency level		
10	Critical thinking	21%	79%
10	Numeracy	22%	78%
9	Adaptive learning	22%	78%
9	Complex problem solving	17%	83%
8	Digital / ICT skills	9%	91%
8	Management	11%	89%
8	Social and interpersonal	20%	80%
7	Verbal communication	27%	73%
5	Creative thinking/design	6%	94%
1	Written communication	100%	0%

■ Step-up to intermediate   ■ Step-up to advanced

ICT = information and communication technology.

Notes: Based on survey of employers in the textile and garment manufacturing industry between June and September 2021 (n=51). Index based on the number of employers indicating a need for workers with basic proficiency to be upskilled for each skill.

Source: Asian Development Bank (Sustainable Development and Climate Change Department).

Skills Supply Trends

Overall, textile and garment manufacturing firms in Uzbekistan appear to face challenges in recruiting well-trained graduates and need to invest substantial resources into training their workers. Less than half of firms surveyed agree that graduates they hired in the past year were adequately prepared for the role by their education or training and close to 80% of firms agree that graduate quality varied significantly between training providers (Figure 13). While 76% of employers said that there are sufficient graduates from relevant training programs to meet their firm's entry-level hiring needs, this is likely due to the investments that employers put into training workers. Of the firms surveyed, 94% indicated that their workers received the appropriate amount and quality of training to do their jobs well and over 70% indicated that they invest sufficiently in worker training (Figure 14).

Figure 13: Employer Sentiment Toward Graduates Hired in the Textile and Garment Manufacturing Industry in Uzbekistan

Close to 80% of employers feel that there is a large of variance in the quality of graduates across different training institutions

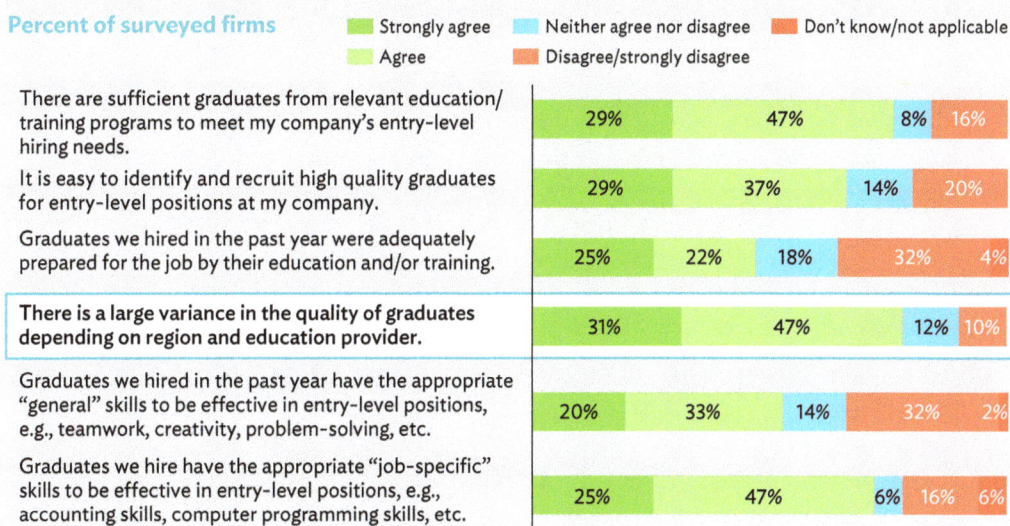

Percent of surveyed firms
- Strongly agree
- Agree
- Neither agree nor disagree
- Disagree/strongly disagree
- Don't know/not applicable

	Strongly agree	Agree	Neither agree nor disagree	Disagree/strongly disagree	Don't know/not applicable
There are sufficient graduates from relevant education/training programs to meet my company's entry-level hiring needs.	29%	47%	8%		16%
It is easy to identify and recruit high quality graduates for entry-level positions at my company.	29%	37%	14%	20%	
Graduates we hired in the past year were adequately prepared for the job by their education and/or training.	25%	22%	18%	32%	4%
There is a large variance in the quality of graduates depending on region and education provider.	31%	47%	12%	10%	
Graduates we hired in the past year have the appropriate "general" skills to be effective in entry-level positions, e.g., teamwork, creativity, problem-solving, etc.	20%	33%	14%	32%	2%
Graduates we hire have the appropriate "job-specific" skills to be effective in entry-level positions, e.g., accounting skills, computer programming skills, etc.	25%	47%	6%	16%	6%

Note: Based on survey of employers in the textile and garment manufacturing industry between June and September 2021 (n=51).
Source: Asian Development Bank (Sustainable Development and Climate Change Department).

Figure 14: Employers' Perception on Training for Employees in the Textile and Garment Manufacturing Industry in Uzbekistan

Close to 70% of employers feel that they currently invest sufficiently in training their employees

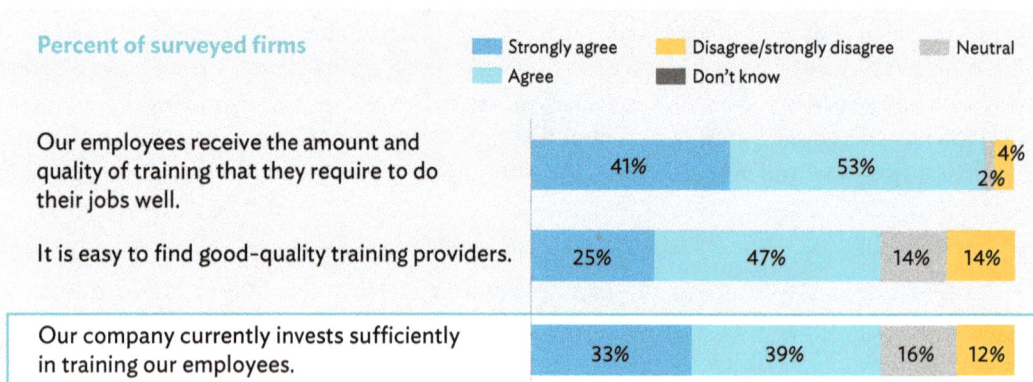

Percent of surveyed firms

	Strongly agree	Disagree/strongly disagree	Neutral
	Agree	Don't know	

Our employees receive the amount and quality of training that they require to do their jobs well.	41%	53%	4% / 2%	
It is easy to find good-quality training providers.	25%	47%	14%	14%
Our company currently invests sufficiently in training our employees.	33%	39%	16%	12%

Note: Based on survey of employers in the textile and garment manufacturing industry between June and September 2021 (n=51).
Source: Asian Development Bank (Sustainable Development and Climate Change Department).

Four types of training channels were examined in relation to how they would be tapped to provide skills training for employees in 2020 and going forward (Table 4).

Table 4: Four Types of Training Channels

	Training Channel	Description
1	On-the-job training	Training that takes place within the firm as the employee performs the actual work. These are typically provided by a more senior or experienced coworker and can also be in the form of internally organized sessions conducted by coworkers.
2	Flexible online training	Online courses that are subsidized or sponsored by the firm for their employees (e.g., courses from online training platforms like Udemy and Coursera). Such online training courses tend to be flexible in terms of when workers may access the content, and typically allow workers to gain industry-recognized micro-credentials or micro-degrees at the end of the course.
3	Professional courses	Short courses that are sponsored or organized by the firm for their employees. These are conducted by professional instructors and are typically held within contained periods spanning at least 1 week and up to 6 months.
4	Formal education courses	Such courses are those taken at higher education or technical and vocational education and training institutions that are subsidized or sponsored by the firm for their employees. Such courses tend to be specially designed for working professionals, such as part-time diplomas or master's degrees.

Source: Asian Development Bank (Sustainable Development and Climate Change Department).

On-the-job training (OJT) is the key training channel used by textile and garment manufacturing firms in Uzbekistan (Figure 15). This is likely due to the use of specialized processes and equipment in the industry so that training is best provided in a hands-on manner. However, more employers expect to use flexible online training channels and professional courses to train workers over the next 5 years. The use of augmented and virtual reality technologies to supplement such training channels could make them more effective. More structured training programs for new workers, particularly those without previous training or experience could also help to raise the overall quality of the textile and garment manufacturing workforce. In particular, it would be critical

Figure 15: Proportion of Employees Receiving Training in 2020 and Requiring Training by 2025 in Each Training Channel in the Textile and Garment Manufacturing Industry in Uzbekistan

On-the-job training is expected to remain the main training channel in Uzbekistan in 5 years' time

Percentage share of employees by training channel

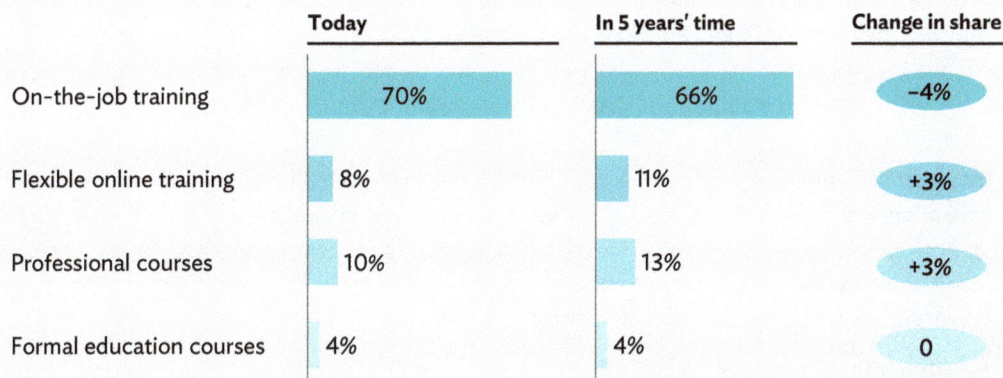

	Today	In 5 years' time	Change in share
On-the-job training	70%	66%	–4%
Flexible online training	8%	11%	+3%
Professional courses	10%	13%	+3%
Formal education courses	4%	4%	0

Notes: Based on survey of employers in the textile and garment manufacturing industry between June and September 2021 (n=51). The sum of all shares for 2020 and by 2025 exceed 100%, as one employee can undergo training in more than one training channel.

Source: Asian Development Bank (Sustainable Development and Climate Change Department).

to provide targeted training to different groups of workers with different capabilities and scope of work (i.e., a manager vs. a factory floor worker, an experienced hire vs. a fresh hire), and training programs would need to be suitably tailored. In Singapore, the Institute of Technical Education (ITE) provides consultancy services to set up structured OJT programs according to the pedagogic competencies and needs of organizations. Organizations that provide training to employees under approved OJT programs are also eligible for government grants.[6]

D.   Construction Industry

Overall, understanding of 4IR technologies and their applications is more limited among construction firms although significant variances are observed in the construction industry. About 35% of the surveyed firms indicated that they had not heard of 4IR before (although the same proportion indicated an advanced understanding of Industry 4.0).

If awareness of 4IR technologies is increased among construction firms in Uzbekistan and full adoption is achieved in the industry, significant net job gains from 4IR can be expected. Around 334,000 new jobs or the equivalent of 25% of the 2020 construction workforce are expected to be created in 2025. This gain is over and beyond the BAU growth of the industry's labor force. The distribution of jobs will also shift slightly, with the share

---

[6]   ITE. Industry Training Schemes. https://www.ite.edu.sg/employers/industry-training-schemes/certified-on-the-job-training-centre.

of technical jobs increasing, although manual jobs will continue to take up the largest proportion of all jobs. In tandem with this shift, digital and/or ICT skills as well as creative thinking and/or design skills will become more important to employers by 2025 according to the employer surveys.

Relevance of Industry 4.0 to the Construction Industry

Research indicates that digital technologies such as building information modeling (BIM), wireless sensors, and additive manufacturing are changing how infrastructure, real estate, and other built assets are designed, constructed, and maintained (Buehler, Buffet, and Castagnino 2018).

Some key 4IR technologies relevant to the construction industry include the following:

(i) **Internet of Things.** Internet of things (IOT) refers to networks of sensors and actuators embedded in machines and other physical objects that connect with one another and the internet. IOT-enabled sensors are also used for smart building management and to identify priorities for maintenance and repair work, as well as collect information on temperature, humidity, noise, and vibration to improve construction worksite safety and efficiency.

(ii) **Additive manufacturing (3D printing).** Additive manufacturing technologies produce physical objects from digital models by adding thin layers of material in succession. 3D printing can be used to create modular parts to use on-site in buildings, including unique architectural features, or even create entire houses. A United States (US) based construction company ICON uses 3D printing technology to produce low-cost housing. ICON brings its printer on-site, squeezing out long tubes of concrete layer by layer, which dry quickly to form the walls of a house thereby eliminating shipping costs in the process. ICON estimates that its 3D printing technology can reduce construction costs by up to 30% and build a house twice as fast as traditional methods (Chea 2021).

(iii) **Augmented reality.** Augmented reality is an interactive experience of a real-world environment where the objects that reside in the real world are enhanced by computer-generated perceptual information, sometimes across multiple sensory modalities. It can be used to showcase 3D models of buildings at a detailed level prior to construction, conduct walk-throughs of the project before execution, as well as reduce errors in the construction process. The start-up XYZ Reality based in the United Kingdom is developing an AR-based helmet that projects detailed holograms of 3D building schematics in front of the user's eyes. It allows construction workers to position objects to a 5-millimeter accuracy and immediately see when something is out of alignment, even to a minute degree. The helmet could potentially save construction firms up to 11% of costs on building projects arising from errors introduced between architectural sketches and finished buildings (Gillet 2021).

(iv) **Cloud computing.** Cloud computing is on-demand access, via the internet, to computing resources, including applications, servers, data storage, development tools, and networking capabilities hosted at a remote data center and managed by a cloud service provider. Cloud computing enables architects and engineers to collaborate on BIMs from different locations, access construction documents, and track project progress remotely and in real time. The use of cloud-based data platforms also enables construction firms to keep information technology (IT) costs low. A 2017 survey conducted in the US showed that nearly **85% of construction contractors used or planned to use cloud-based solutions** (Built Worlds 2017). Research by BCG Platinion estimates that cloud computing can reduce overall IT spend by as much as 10% overall (Google Cloud Blog 2021).

(v) **Cybersecurity.** Cybersecurity is the protection of internet-connected systems from unauthorized exploitation. As the construction industry adopts more digital solutions, the risks of cyberattacks are amplified with vast amounts of confidential information digitally stored and shared. Cybersecurity technologies are needed to protect proprietary information such as blueprints and bidding data, and to avoid downtime and property damage. The use of 4IR technologies can reduce the impact of security breaches on firms. IBM's research estimates that organizations with fully deployed security AI and automation experienced breach costs of $2.9 million on average, compared to $6.71 million at organizations that did not deploy such technologies (IBM Security 2021).

(vi) **Digital twin.** A digital twin is a virtual representation of an object or system that spans its life cycle. It is updated using real-time data and uses simulation and machine learning to aid decision-making (Armstrong 2020). In the construction and real estate industry, digital twin technologies allow different aspects of a worksite, building, or asset to be analyzed quickly to identify problems and optimize performance (Lawton 2021). Research by Ernst & Young estimates that the adoption of digital twin can reduce real estate operating costs by up to 35%. It can also decrease carbon emissions, deliver a healthier workplace, and enhance user experiences (Techwire Asia 2021).

For the employer survey, 51 construction firms in Uzbekistan were surveyed. Firms in the construction industry vary widely in their understanding of 4IR. Around half of firms surveyed indicated a limited understanding of 4IR technologies and their applications. However, 35% said that they had an advanced understanding of such technologies (Figure 16). In addition, 59% of construction firms indicated that they expect firms in their supply chain (e.g., material suppliers, realtors, equipment providers) to have a strong understanding of 4IR technologies and their applications (Figure 17). These findings combined with expert interviews and consultations with key stakeholders in Uzbekistan suggest that understanding and deployment of 4IR technologies is relatively strong in the construction and real estate industry in Uzbekistan. However, the usage of such technologies is generally limited to larger and more mature firms involved in large-scale infrastructure projects that require such

Figure 16: Understanding of Industry 4.0 Technologies in the Construction Industry in Uzbekistan

Half of the firms surveyed in the construction industry have limited understanding of 4IR technologies and their applications

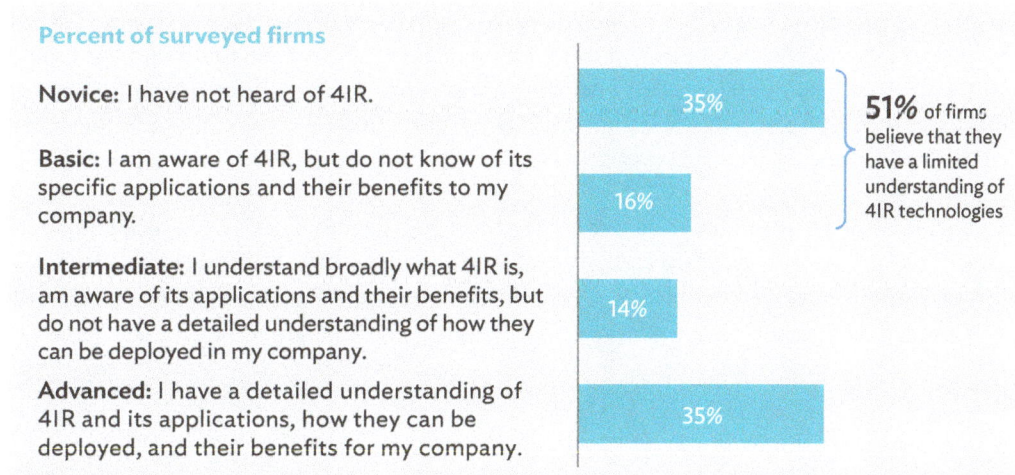

Percent of surveyed firms

Novice: I have not heard of 4IR.   35%

Basic: I am aware of 4IR, but do not know of its specific applications and their benefits to my company.   16%

Intermediate: I understand broadly what 4IR is, am aware of its applications and their benefits, but do not have a detailed understanding of how they can be deployed in my company.   14%

Advanced: I have a detailed understanding of 4IR and its applications, how they can be deployed, and their benefits for my company.   35%

51% of firms believe that they have a limited understanding of 4IR technologies

4IR = Fourth Industrial Revolution.
Note: Based on survey of employers in the construction industry between June and September 2021 (n=51).
Source: Asian Development Bank (Sustainable Development and Climate Change Department).

Figure 17: Understanding of Industry 4.0 Technologies Among Firms in the Construction Supply Chain in Uzbekistan

Fifty-nine percent of construction firms believe that companies in their supply chain have a good understanding of 4IR technologies

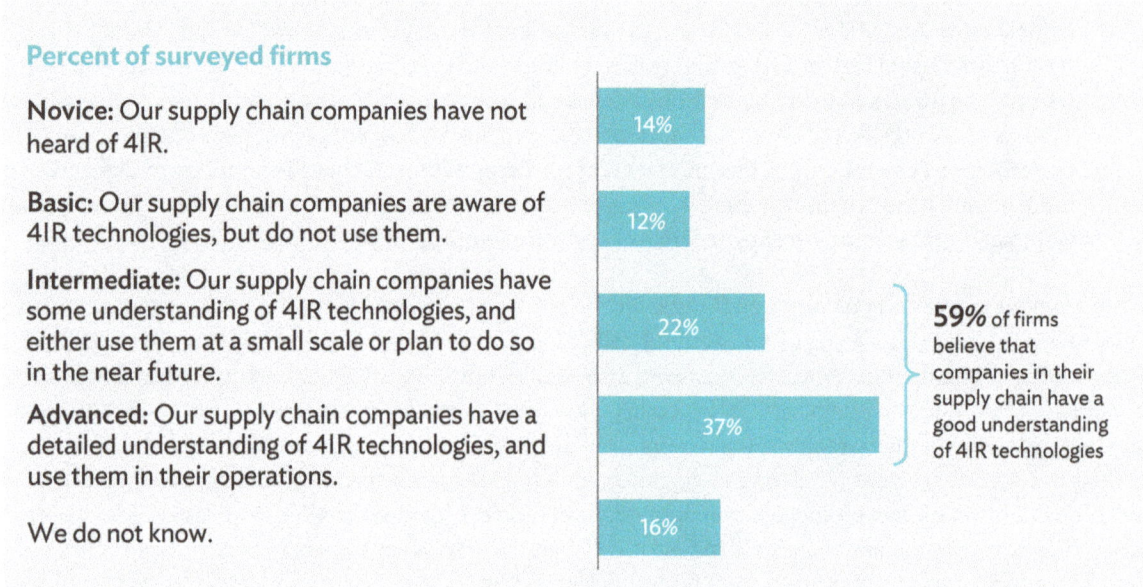

Percent of surveyed firms

Novice: Our supply chain companies have not heard of 4IR.

14%

Basic: Our supply chain companies are aware of 4IR technologies, but do not use them.

12%

Intermediate: Our supply chain companies have some understanding of 4IR technologies, and either use them at a small scale or plan to do so in the near future.

22%

Advanced: Our supply chain companies have a detailed understanding of 4IR technologies, and use them in their operations.

37%

We do not know.

16%

59% of firms believe that companies in their supply chain have a good understanding of 4IR technologies

4IR = Fourth Industrial Revolution.
Note: Based on survey of employers in the construction industry between June and September 2021 (n=51).
Source: Asian Development Bank (Sustainable Development and Climate Change Department).

technologies, while smaller firms (e.g., subcontractors) involved in smaller-scale projects see less need for such technologies and lack the resources to deploy them. In addition, most advanced technologies and systems are imported as domestic innovation is at a fledging stage. This means that firms need to expend resources to adapt these systems to local conditions and train their workers to use them, placing smaller firms with less resources at a further disadvantage. Given that SMEs in Uzbekistan employ close to 80% of the workforce, targeted programs to support SMEs to adopt digital tools could help to ensure that gains from 4IR are equitably distributed.

As with firms in the textile and garment manufacturing industry, construction firms have strong expectations of the potential labor productivity gains from adopting 4IR technologies. On average, firms expect output per worker to increase by 60% in 2025 with the adoption of 4IR technologies (Figure 18). Past research further estimates that full-scale digitization could generate an estimated 12%–20% in annual cost savings for the construction industry. It is critical to build awareness of these benefits among construction firms in Uzbekistan to ensure that these benefits can be reaped (Buehler, Buffet, and Castagnino 2018).

Currently, IOT and cybersecurity are the key 4IR technologies deployed by construction firms in Uzbekistan. However, adoption rates for other technologies, particularly augmented reality technologies, are expected to rise between 2020 and 2025 (Figure 19). As of 2020, 20% of construction firms surveyed deploy IOT technologies across all possible functions. There are a range of applications for IOT technologies in the construction industry.

Figure 18: Expected Increase in Output per Worker Due to Industry 4.0 Technologies in the Construction Industry in Uzbekistan, 2020–2025

Over half of construction firms expect 4IR technologies to increase output per worker by more than 50% in 5 years' time

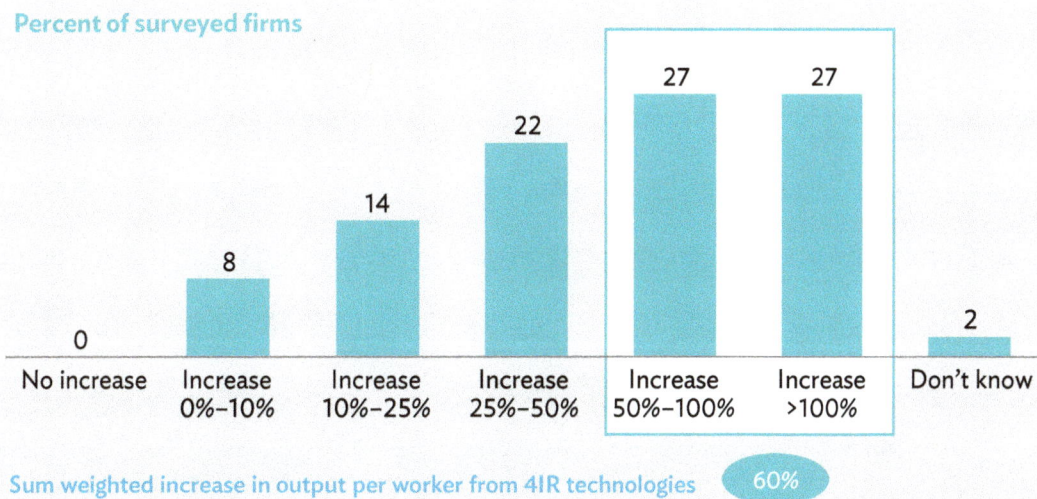

4IR = Fourth Industrial Revolution.

Notes: Based on survey of employers in the construction industry between June and September 2021 (n=51). Calculated using sum-weighted average of output increase by the number of firms indicating different levels of expected increase in output, i.e., 0%, 0%–10%, 10%–25%, 25%–50%, 50%–100%, and over 100%. The midpoint of the range for each option for expected increase in output is used. For expected output increase of over 100%, the lower bound of 100% is used.

Source: Asian Development Bank (Sustainable Development and Climate Change Department).

IOT-enabled sensors can collect information on temperature, humidity, noise, and vibration to improve construction worksite safety and efficiency. Triax Technologies has a suite of wearable solutions that improve worksite safety by allowing firms to control site entry at hazardous sites and providing real-time preventive alerts and operational data insights among other functions (Triax 2021). Of the construction firms surveyed, 41% deploy cybersecurity technologies across all functions. With the expected rise in use of digital technologies and IOT-enabled devices on the worksite, cybersecurity technologies will become more critical to protect digital systems as well as information such as blueprints and bidding data, and to avoid downtime and property damage.

Construction firms in Uzbekistan also recognize the potential of augmented reality technologies, with 90% of the firms expect to adopt such technologies to some extent by 2025. Augmented reality can be used to allow various teams involved in a construction project to meet virtually for discussions, and are particularly useful against the backdrop of physical travel restrictions imposed against the backdrop of the COVID-19 pandemic. For instance, Suffolk Construction uses virtual reality (VR) technology to enable its engineering teams to meet virtually to coordinate, plan, and resolve issues, irrespective of their geographical locations. Users can join discussions from their desktops, wearing a VR headset, to review project designs, spot issues, and make changes, all inside a virtual environment (FutureCIO 2020).

Figure 19: Current and Future Adoption of Relevant Industry 4.0 Technologies in the Construction Industry in Uzbekistan

Internet of Things and cybersecurity are expected to be particularly crucial for the construction industry in the next 5 years

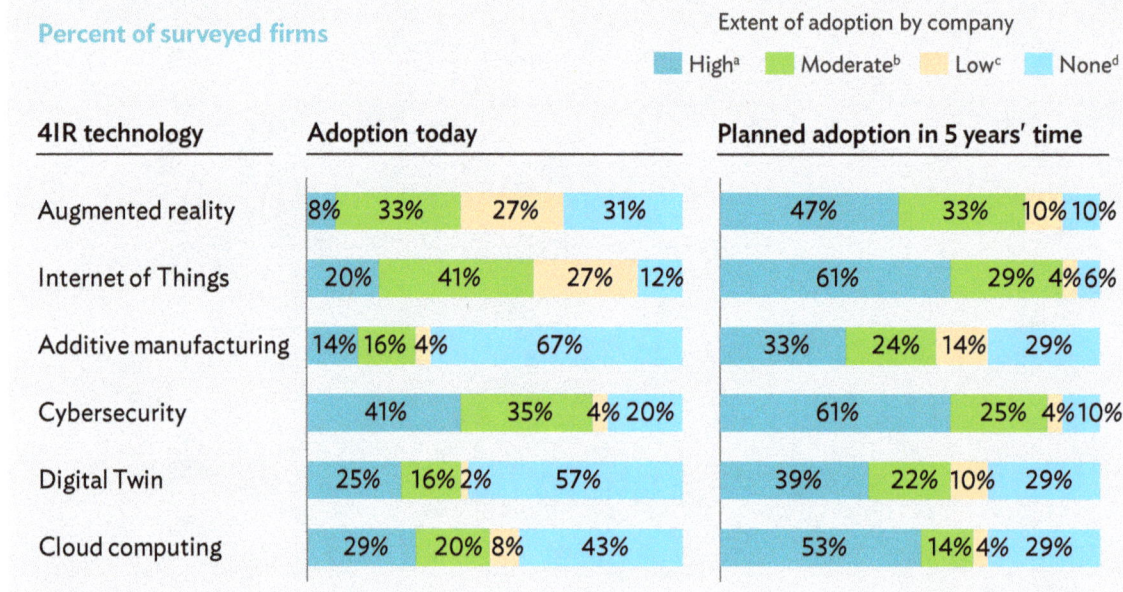

Percent of surveyed firms

Extent of adoption by company

■ High[a]   ■ Moderate[b]   ■ Low[c]   ■ None[d]

4IR technology	Adoption today				Planned adoption in 5 years' time			
Augmented reality	8%	33%	27%	31%	47%	33%	10%	10%
Internet of Things	20%	41%	27%	12%	61%	29%	4%	6%
Additive manufacturing	14%	16%	4%	67%	33%	24%	14%	29%
Cybersecurity	41%	35%	4%	20%	61%	25%	4%	10%
Digital Twin	25%	16%	2%	57%	39%	22%	10%	29%
Cloud computing	29%	20%	8%	43%	53%	14%	4%	29%

4IR = Fourth Industrial Revolution.

[a] "High": Firm has fully deployed the technology across all possible functions in the enterprise and/or has plans to fully deploy the technology across all possible functions in the future.

[b] "Moderate": Firm has implemented the technology, but not fully deployed across all possible functions in the enterprise and/or plans to implement the technology across a few functions in the future.

[c] "Low": Firm is experimenting with the technology at a very limited scale within the enterprise and/or plans to experiment with the technology in the future.

[d] "None": Firm has not used technology at all within the enterprise and/or has no plans to use the technology in the future.

Note: Based on survey of employers in the construction industry between June and September 2021 (n=51).

Source: Asian Development Bank (Sustainable Development and Climate Change Department).

Less than 50% of construction firms surveyed believed that the COVID-19 pandemic will accelerate the use of 4IR technologies and close to a third disagree that COVID-19 will accelerate the use of such technologies (Figure 20). As in textile and garment manufacturing firms, these muted expectations could be due to the general global economic uncertainty created by the COVID-19 pandemic, so that firms are unwilling to invest significant resources in new technologies.

Figure 20: Perception on the Impact of the COVID-19 Pandemic on the Adoption of Industry 4.0 Technologies in the Construction Industry in Uzbekistan

Less than half of employers believe that the COVID-19 pandemic has accelerated or will accelerate the use 4IR technologies

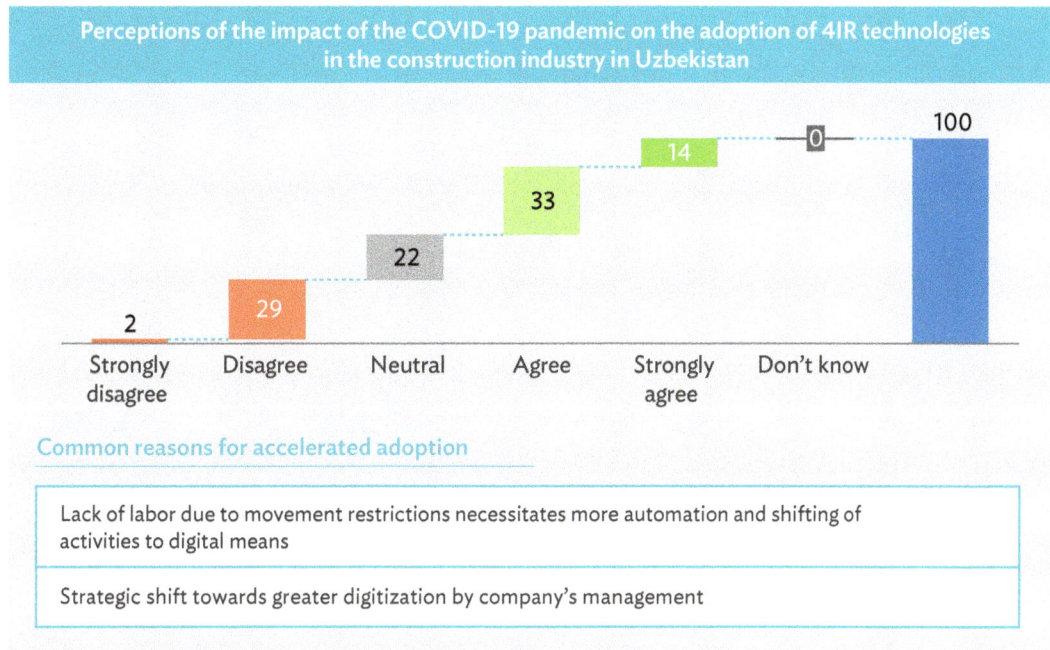

Perceptions of the impact of the COVID-19 pandemic on the adoption of 4IR technologies in the construction industry in Uzbekistan

Strongly disagree	Disagree	Neutral	Agree	Strongly agree	Don't know	
2	29	22	33	14	0	100

Common reasons for accelerated adoption

Lack of labor due to movement restrictions necessitates more automation and shifting of activities to digital means

Strategic shift towards greater digitization by company's management

4IR = Fourth Industrial Revolution, COVID=19 = coronavirus disease.
Note: Based on survey of employers in the construction industry between June and September 2021 (n=51).
Source: Asian Development Bank (Sustainable Development and Climate Change Department).

Skills Demand Analysis

Job Implications

The full adoption of 4IR technologies by firms in Uzbekistan is expected to create over 334,000 new jobs or the equivalent of 25% of the 2020 construction workforce, over and beyond the BAU growth of the industry's labor force by 2025. This is as the productivity effect again outweighs the displacement effect. Job displacement due to automation is expected to displace around 556,000 positions but create around 890,000 new jobs, over and beyond BAU growth of the industry's labor force (Figure 21).

As in the textile and garment manufacturing industry, more construction jobs will be created in Uzbekistan by 2025 even without 4IR. The construction workforce grew at approximately 2.3% per annum[7] from 2015 and 2020. If this growth were extrapolated up to 2025, an additional 160,000 jobs could have been created even without adopting 4IR technologies. However, the full adoption of 4IR technologies could see a total of around 500,000 more construction jobs in 2025 than in 2020.

---

[7]   Calculated using industry and workforce data from State Committee of the Republic of Uzbekistan on Statistics.

Figure 21: Estimated impact of Industry 4.0 on Number of Jobs in the Construction Industry in Uzbekistan, 2020–2025

Less than half of employers believe that the COVID-19 pandemic has accelerated or will accelerate the use 4IR technologies

Percent of jobs impacted due to displacement and productivity effects of 4IR in 5 years' time

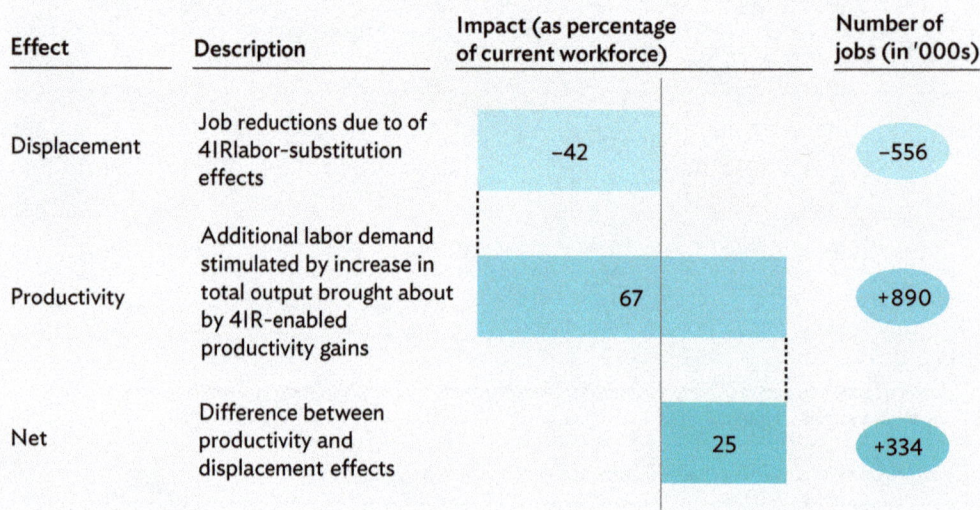

Effect	Description	Impact (as percentage of current workforce)	Number of jobs (in '000s)
Displacement	Job reductions due to of 4IRlabor-substitution effects	−42	−556
Productivity	Additional labor demand stimulated by increase in total output brought about by 4IR-enabled productivity gains	67	+890
Net	Difference between productivity and displacement effects	25	+334

4IR = Fourth Industrial Revolution.

Notes: Based on survey of employers in the construction industry between June and September 2021 (n=51). Industry employment data is from State Committee of the Republic of Uzbekistan on Statistics. Labor Market. https://stat.uz/en/official-statistics/labor-market and output data is from State Committee of the Republic of Uzbekistan on Statistics. Construction. https://stat.uz/en/official-statistics/construction.

Source: Asian Development Bank (Sustainable Development and Climate Change Department).

As in the textile and garment manufacturing industry, the impact on jobs in the construction industry will differ by occupation. This report categorized jobs in Uzbekistan's construction industry into five occupational groups (see Table 5).

Table 5: Occupational Groups in the Construction Industry

	Occupational Group	Examples of Job Titles
1	**Technical**	• Building engineer • Architect
2	**Managerial**	• Chief executive officer • Supervisor
3	**Customer-facing**	• Hotline operator • Business development manager
4	**Administrative**	• Secretary • Finance executive
5	**Elementary and/or manual jobs**	• Forklift operator • Construction site worker

Source: Asian Development Bank and AlphaBeta.

Box 2: The Potential Cost of Government Inaction on Job Creation under Industry 4.0 Adoption

The net gains estimated in this report assumes that the entire textile and garment manufacturing and construction industries adopt Fourth Industrial Revolution (4IR) technologies. For this to occur, the Government of Uzbekistan has a crucial role to play in ensuring that appropriate policies are in place to encourage participation, especially for firms that face significant adoption barriers. Effective policies are also needed to smooth labor market frictions, where public reskilling initiatives could be necessary for displaced workers to successfully transition into newly created jobs. An interesting analysis may be to understand how the breadth and effectiveness of policies impact net job creation under 4IR technology adoption. This will require empirical evaluations (e.g., simulations of impact on labor productivity growth under different policy scenarios) that are beyond the scope of this report and could be the subject of future research.

A simple thought exercise to provide some immediate guidance could be to assume that only the proportion of firms (71% in the textile and garment manufacturing industry and 49% in the construction industry) that indicated an intermediate or advanced understanding of 4IR technologies and their applications in 2020, would reap the productivity gains of 4IR. In this case, only 71% of the 100,000 net job gains (around 71,000 jobs) expected in the full adoption scenario will be realized for the textile and garment manufacturing industry. Similarly, in the construction industry, only 49% of the 334,000 jobs (around 163,700 jobs) expected to be created in the full adoption scenario will be realized.

This research further demonstrates that the impact of 4IR goes beyond job creation. 4IR will change the nature of jobs in the textile and garment manufacturing and construction industries, and the types of skills that will be needed to take on these jobs. Studies have demonstrated that this will free up time for workers doing routine tasks to take on higher-value tasks such as creative work. This would increase job satisfaction among workers and potentially lead to higher wages (AlphaBeta 2017).

Source: AlphaBeta. 2017. *The Automation Advantage.* https://alphabeta.com/wp-content/uploads/2017/08/The-Automation-Advantage.pdf.

Similar to the textile and garment manufacturing industry, employers expect the number of jobs across most occupational groups to increase by 2025 (Figure 22). This is likely due to (i) a significant proportion of firms already adopting 4IR technologies in some or all functions so that displacement is expected to be limited, and (ii) expectations that the construction industry will continue to grow as Uzbekistan develops its urban infrastructure. Notwithstanding this, there is a clear trend toward more technical roles in the industry with the adoption of 4IR technologies. The sharpest increase is expected in technical roles with three-quarters of employers expecting to see the number of technical workers increase.

The distribution of jobs in the construction industry will see some shifts with the adoption of 4IR technologies. Uzbekistan's construction firms expect that by 2025 , the proportion of technical workers will increase although manual workers will continue to form the largest group of workers in the industry (Figure 23). The high proportion of manual jobs could be due to current technological and resource constraints in automating construction processes. Compared to manufacturing processes that are repetitive and can take place in a controlled environment, construction tasks are more difficult to automate, and machines would need to be able to adapt to real-time changes in their environment including temperature and humidity changes with little programming (Association for Advancing Automation 2018). As such, it is unlikely for 4IR technologies to reduce the number of manual workers substantially in the near term. That said, these technologies can still simplify construction tasks and improve workplace safety for workers, improving the quality of jobs in other ways. For instance, Intellinium's IOT-enabled smart shoes allow workers to send and receive messages hands-free. During an emergency, a worker can acknowledge receiving a message to evacuate by touching a toe to the sensor or quickly

Figure 22: Expected Impact of Industry 4.0 on the Number of Jobs between 2020 and 2025 in the Construction Industry in Uzbekistan

In the construction industry, most firms expect technical jobs to increase due to the adoption of 4IR

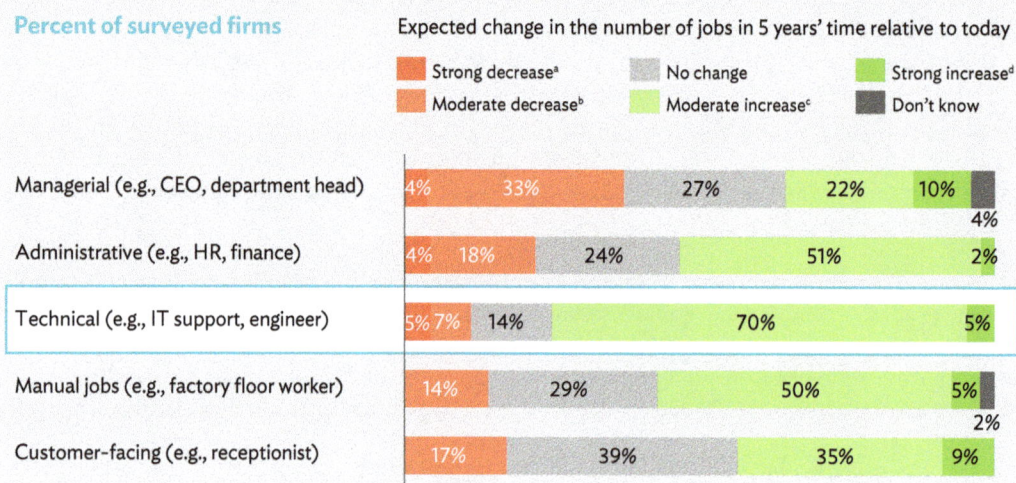

Percent of surveyed firms

Expected change in the number of jobs in 5 years' time relative to today

- Strong decrease[a]
- Moderate decrease[b]
- No change
- Moderate increase[c]
- Strong increase[d]
- Don't know

Job category	Strong decrease	Moderate decrease	No change	Moderate increase	Strong increase	Don't know
Managerial (e.g., CEO, department head)	4%	33%	27%	22%	10%	4%
Administrative (e.g., HR, finance)	4%	18%	24%	51%		2%
Technical (e.g., IT support, engineer)	5%	7%	14%	70%		5%
Manual jobs (e.g., factory floor worker)		14%	29%	50%	5%	2%
Customer-facing (e.g., receptionist)		17%	39%	35%		9%

CEO = chief executive officer, HR = human resources, IT = information technology.

a   Greater than or equal to 50% decrease in number of jobs.
b   Less than 50% decrease in number of jobs.
c   Less than 50% increase in number of jobs.
d   Greater than or equal to 50% increase in number of jobs.

Note: Based on survey of employers in the construction industry between June and September 2021 (n=51).

Source: Asian Development Bank (Sustainable Development and Climate Change Department).

send an alert if he or she is injured. Built-in sensors detect falls or shocks, immediately notifying others of the worker's location to speed up response time to occupational accidents.[8]

In addition, the proportion of managerial workers is also expected to fall with the adoption of 4IR technologies according to the employer surveys. This could be due to the adoption of 4IR technologies such as IOT and digital twin technologies that allow managers such as construction foreman or worksite managers to more easily track the deployment of manpower and equipment so that fewer managers are required.

Unlike the textile and garment manufacturing industry, a larger proportion of the job gains will go to male workers compared to female workers (Figure 24). This is due to the relatively higher proportion of male workers in the construction industry traditionally, due to the physical demands of most roles. However, as 4IR technologies change the nature of jobs and tasks in the industry, there is opportunity for more female workers to join the construction workforce.

---

8   Sierra Wireless. Intellinium selects Sierra Wireless' device-to-cloud IOT solution for industry's first smart safety shoe to protect workers. https://www.sierrawireless.com/company/newsroom/pressreleases/2018/02/intellinium-selects-sierra-wireless-device-to-cloud-iot-solution-for-smart-safety-shoe/.

Figure 23: Composition of Jobs in 2020 and by 2025 by Occupational Group in the Construction Industry in Uzbekistan

The distribution of jobs will change with technical jobs seeing the largest increase in 5 years' time

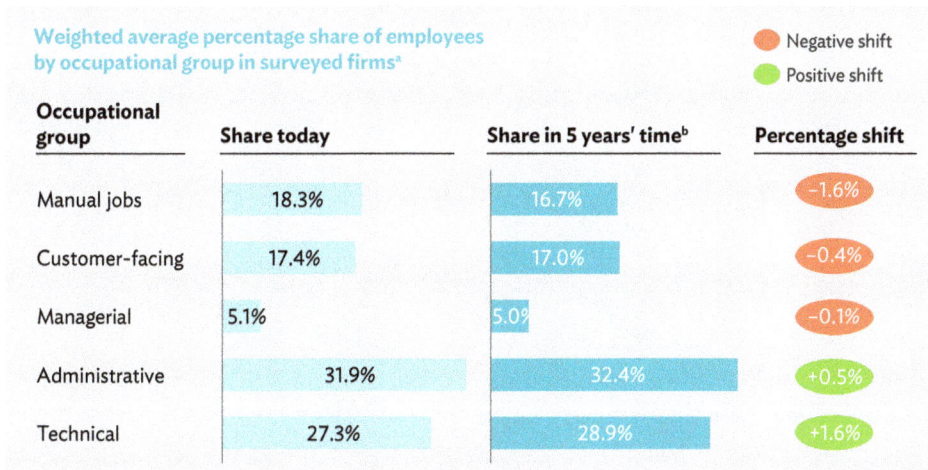

Weighted average percentage share of employees by occupational group in surveyed firms[a]

● Negative shift
● Positive shift

Occupational group	Share today	Share in 5 years' time[b]	Percentage shift
Manual jobs	18.3%	16.7%	−1.6%
Customer-facing	17.4%	17.0%	−0.4%
Managerial	5.1%	5.0%	−0.1%
Administrative	31.9%	32.4%	+0.5%
Technical	27.3%	28.9%	+1.6%

[a] Average share of employees in surveyed firms is weighted by the number of employees in each firm, as indicated by respondents. Percentages may not add up to 100% due to rounding.

[b] The change in the number of workers in each job type is based on the number of firms indicating different levels of changes in number of jobs, i.e., "strong increase," "moderate increase," "no change," "moderate decrease," "strong decrease." The midpoint of the range for each option for expected change is used. For expected increase or decrease of over 50%, we used the low-bound of 50% was used.

Note: Based on survey of employers in the construction industry between June and September 2021 (n=51).

Source: Asian Development Bank (Sustainable Development and Climate Change Department).

Figure 24: Estimated Net Job Gains by Gender from Industry 4.0 Adoption Between 2020 and 2025 in the Construction Industry in Uzbekistan

Targeted programs could increase female participation in the construction industry

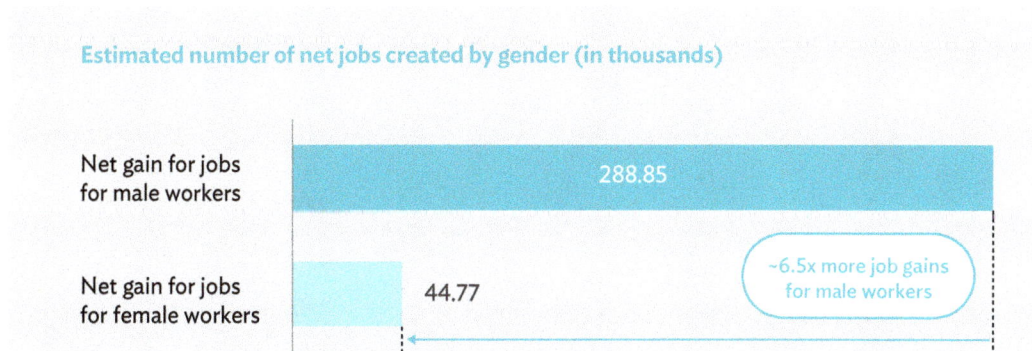

Estimated number of net jobs created by gender (in thousands)

Net gain for jobs for male workers	288.85
Net gain for jobs for female workers	44.77

~6.5x more job gains for male workers

Notes: Based on survey of employers in the construction industry between June and September 2021 (n=51). Industry employment data is from State Committee of the Republic of Uzbekistan on Statistics. Labor Market. https://stat.uz/en/official-statistics/labor-market.

Source: Asian Development Bank (Sustainable Development and Climate Change Department).

Task Implications

The employer survey reveals a sharp drop in the proportion of time that employers expect workers to spend on routine tasks, both physical and interpersonal, by 2025, with the adoption of 4IR technologies (Figure 25). While automation is unlikely to reduce the number of human workers on the worksite significantly in the near term, some technologies can reduce the amount of time workers spent on routine tasks. For instance, 3D printing can reduce time spent on routine, physical tasks such as bricklaying and precast production. In Singapore, trials showed that building a 3D-printed room, including the manual insertion of steel reinforcement bars into the structure, and fitting in windows and a door, took about 6 days. In contrast, it would take more than 2 months to build a similar room using the conventional method of precast production (*The Straits Times* 2019). Routine, interpersonal tasks such as tracking the location of site workers and deploying them from one part of the site to another can also be simplified through using IOT-enabled devices to track worker location and ensure their safety. Instead, workers will spend more time on nonroutine and analytical tasks as they would need to adjust the location of IOT devices and sensors; or operate, and maintain advanced machinery and digital platforms.

Figure 25: Time Spent by Employees on Tasks at Work in 2020 and by 2025 in the Construction Industry in Uzbekistan

Adoption of 4IR technologies could shift the overall distribution of weekly working hours toward analytical and nonroutine tasks

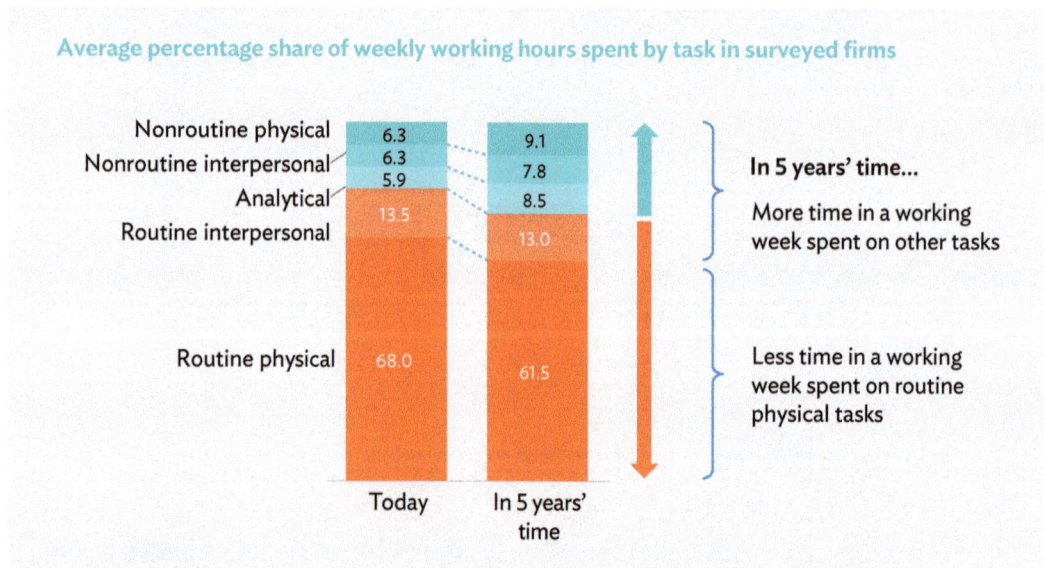

Average percentage share of weekly working hours spent by task in surveyed firms

	Today	In 5 years' time
Nonroutine physical	6.3	9.1
Nonroutine interpersonal	6.3	7.8
Analytical	5.9	8.5
Routine interpersonal	13.5	13.0
Routine physical	68.0	61.5

In 5 years' time...
More time in a working week spent on other tasks

Less time in a working week spent on routine physical tasks

4IR = Fourth Industrial Revolution.
Note: Based on survey of employers in the construction industry between June and September 2021 (n=51). Figures include rounding adjustments.
Source: Asian Development Bank (Sustainable Development and Climate Change Department).

Among the employers that feel that a step-up from basic proficiency in critical thinking skills is needed, 37% feel that a step-up to the intermediate level of proficiency is needed while 63% would like to see a step-up to advanced proficiency. In contrast, while digital and/or ICT skills will be increasingly valued by employers by 2025, most employers assessed the current proficiency levels of employees in this area to be sufficient.

Skills Supply Trends

Like the textile and garment manufacturing industry, some employers in the construction industry find it difficult to identify and hire graduates that are sufficiently prepared for the job by their education or training

Figure 26: Importance of Skills in 2020 and for Industry 4.0 Adoption by 2025 in the Construction Industry in Uzbekistan

Creative thinking/design will be the most important skill for 4IR technology adoption in industry in 5 years' time

Importance ranking	Today[a]	In 5 years' time[b]	Change in ranking
1	Written communication	Creative thinking/design	+6
2	Critical thinking	Digital/ICT skills	+1
3	Digital/ICT skills	Critical thinking	-1
4	Social and interpersonal	Numeracy	+1
5	Numeracy	Adaptive learning	+5
6	Management	Complex problem solving	+2
7	Creative thinking/design	Written communication	-6
8	Complex problem solving	Management	-2
9	Verbal communication	Verbal communication	-
10	Adaptive learning	Social and interpersonal	-6

Skills of increasing importance in 5 years' time   Skills of decreasing importance in 5 years' time   Skills with no change in importance in 5 years' time

ICT = information and communication technology.
[a] Evaluated using employer survey and supported by job portal data.
[b] Evaluated using the employer survey.
Notes: Based on survey of employers in the construction industry between June and September 2021 (n=51); and job data on the IT–BPO industry from the job portal HeadHunter Uzbekistan. Figures include rounding adjustments.
Source: Asian Development Bank (Sustainable Development and Climate Change Department).

Skills Implications

The shift toward more analytical tasks is also reflected in the changes in the skills valued by employers. Creative thinking and design skills as well digital and/or ICT skills will be increasingly sought after by employers by 2025 (Figure 26). The ability of 4IR technologies, particularly IOT technologies, to reduce human interaction in the construction industry is also reflected in the findings. Skills such as management, verbal communication, and social and interpersonal skills will drop significantly in relative importance. These skills are particularly relevant for managerial job roles and their decline in importance is consistent with the shift in workforce distribution away from managerial occupations with the adoption of 4IR technologies. Comparing the skills that are prioritized by employers for 4IR adoption in 2025 against skills in which a step-up in proficiency is needed, our analysis shows that workers would require a significant step-up in critical thinking skills between 2020 and 2025 (Figure 27).

Figure 27: Required Step-Up in Employee Proficiency Level from Today for Industry 4.0 Adoption in the Construction Industry in Uzbekistan, 2020–2025

To be 4IR-ready, workers would require significant proficiency leaps in written and verbal communication skills

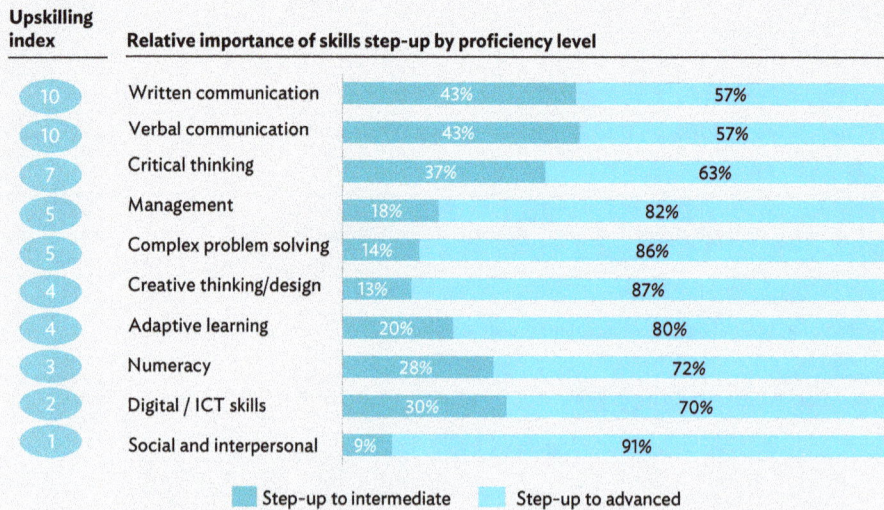

Upskilling index	Relative importance of skills step-up by proficiency level	Step-up to intermediate	Step-up to advanced
10	Written communication	43%	57%
10	Verbal communication	43%	57%
7	Critical thinking	37%	63%
5	Management	18%	82%
5	Complex problem solving	14%	86%
4	Creative thinking/design	13%	87%
4	Adaptive learning	20%	80%
3	Numeracy	28%	72%
2	Digital / ICT skills	30%	70%
1	Social and interpersonal	9%	91%

4IR = Fourth Industrial Revolution, ICT = information and communication technology.

Note: Based on survey of employers in the construction industry between June and September 2021 (n=51). Index is based on the number of employers indicating a need for workers with basic proficiency to be upskilled for each skill.

Source: Asian Development Bank (Sustainable Development and Climate Change Department).

Figure 28: Employer Sentiment Toward Graduates Hired in the Construction Industry in Uzbekistan

Over 70% of employers surveyed observe large variance in the quality of graduates across different traning institutions

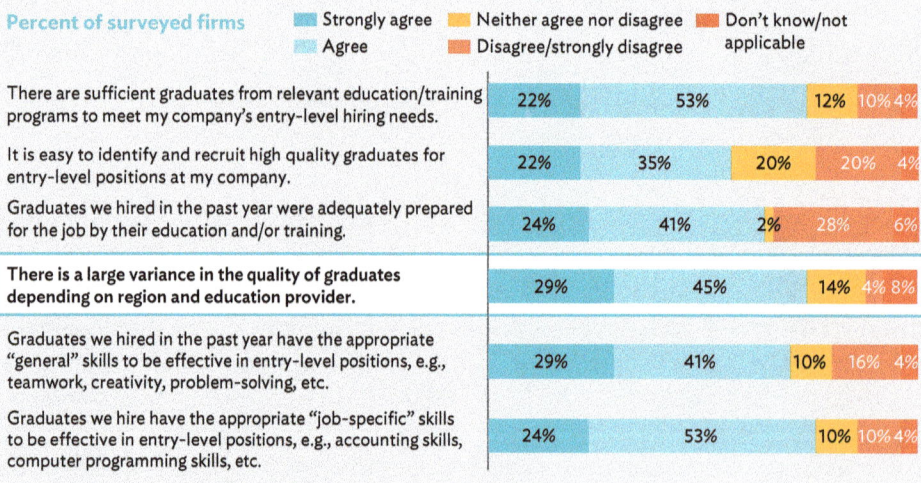

Percent of surveyed firms	Strongly agree	Agree	Neither agree nor disagree	Disagree/strongly disagree	Don't know/not applicable
There are sufficient graduates from relevant education/training programs to meet my company's entry-level hiring needs.	22%	53%	12%	10%	4%
It is easy to identify and recruit high quality graduates for entry-level positions at my company.	22%	35%	20%	20%	4%
Graduates we hired in the past year were adequately prepared for the job by their education and/or training.	24%	41%	2%	28%	6%
There is a large variance in the quality of graduates depending on region and education provider.	29%	45%	14%	4%	8%
Graduates we hired in the past year have the appropriate "general" skills to be effective in entry-level positions, e.g., teamwork, creativity, problem-solving, etc.	29%	41%	10%	16%	4%
Graduates we hire have the appropriate "job-specific" skills to be effective in entry-level positions, e.g., accounting skills, computer programming skills, etc.	24%	53%	10%	10%	4%

Note: Based on survey of employers in the construction industry between June and September 2021 (n=51).

Source: Asian Development Bank (Sustainable Development and Climate Change Department).

(Figure 28). A majority also agree that there is significant variance in the quality of graduates from different training institutions. Despite the difficulties in recruiting good quality candidates, it appears that most employers are willing to provide training. Over 90% of employers agree that employees receive sufficient training to do their jobs well, and more than 70% of firms surveyed said they currently invest sufficiently in training their workers (Figure 29). Programs to enable the construction industry to work more closely with training institutions to provide industry-relevant training could help to improve the readiness of graduates for the workforce.

Like the textile and garment manufacturing industry, OJT training is currently the most common training channel for firms in the construction industry, with firms providing 72% of employees with such training. This suggests that training institutions are not able to sufficiently prepare their graduates for the workforce, and employers need to invest substantial resources into filling the gap. Employers expect to see some shifts in the mode of training in by 2025, with the use of online training channels and professional courses—short-term courses lasting less than 6 months—expected to increase in the next 5 years. As with the textile and garment manufacturing industry, the use of 4IR technologies, such as augmented reality technologies or simulators, could help to make such training approaches more effective. VR-based training is particularly relevant for the construction as untrained workers face a high risk of accidents. Researchers at the University of New South Wales in Australia developed a virtual reality platform that allows construction workers to navigate life-threatening scenarios using a computer or virtual reality headset (Global Construction Review 2016). Virtual reality headsets are also used in Singapore to allow construction workers to experience the danger of working at heights and train them to enforce good safety habits in a safe environment. Workers undergoing training can see the consequences of unsafe practices, such as the lack of guardrails at a site, without being physically endangered (*The Straits Times* 2019b).

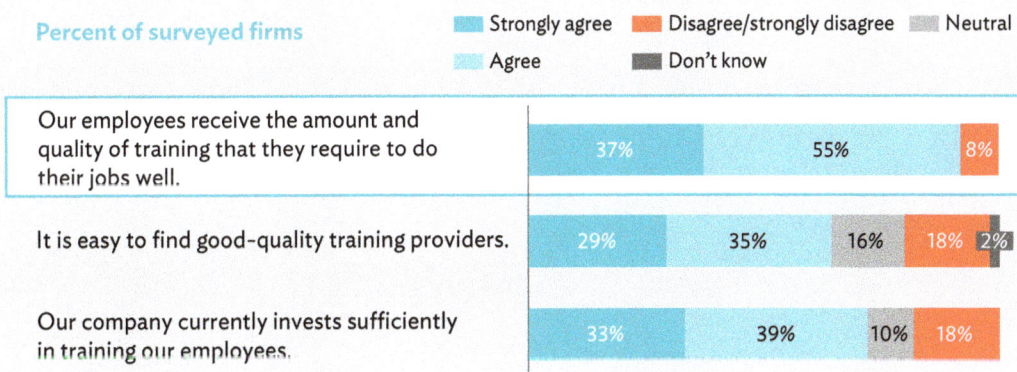

Figure 29: Employers' Perception on Training for Employees in the Construction Industry in Uzbekistan

More than 90% of employers reported that employees receive sufficient training to do their jobs well

Percent of surveyed firms

Legend: Strongly agree | Disagree/strongly disagree | Neutral | Agree | Don't know

	Strongly agree	Agree	Neutral	Disagree/strongly disagree	Don't know
Our employees receive the amount and quality of training that they require to do their jobs well.	37%	55%		8%	
It is easy to find good-quality training providers.	29%	35%	16%	18%	2%
Our company currently invests sufficiently in training our employees.	33%	39%	10%	18%	

Note: Based on survey of employers in the construction industry between June and September 2021 (n=51).
Source: Asian Development Bank (Sustainable Development and Climate Change Department).

Figure 30: Proportion of Employees Receiving Training in 2020 and Requiring Training by 2025 for the Construction Industry in Uzbekistan

The proportion of workers undergoing flexible online training and/or professional courses is expected to increase most in 5 years' time

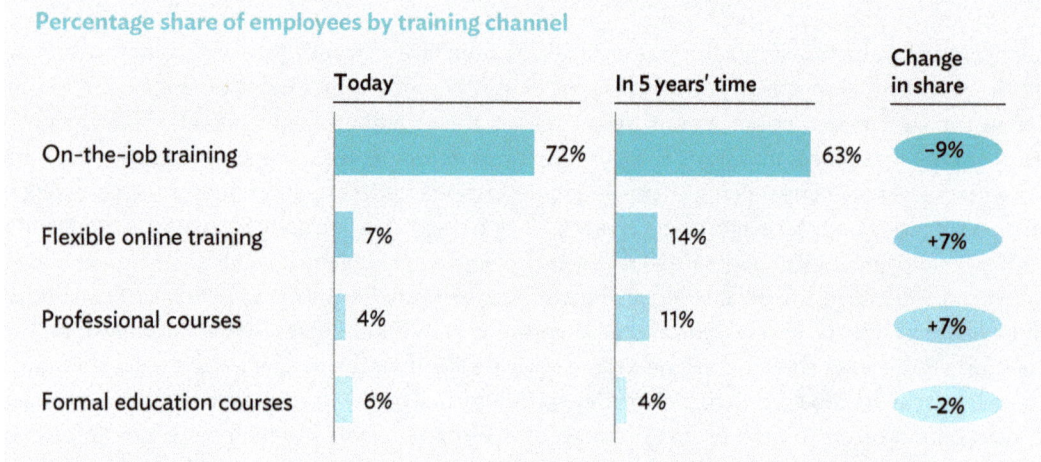

Percentage share of employees by training channel

	Today	In 5 years' time	Change in share
On-the-job training	72%	63%	−9%
Flexible online training	7%	14%	+7%
Professional courses	4%	11%	+7%
Formal education courses	6%	4%	-2%

Notes: Based on survey of employers in the construction industry between June and September 2021 (n=51). The sum of all shares for today and in 5 years exceeds 100%, as one employee can undergo training in more than one training channel.
Source: Asian Development Bank (Sustainable Development and Climate Change Department).

E.   Emerging Jobs

Through the employer surveys conducted and scrapping of online job portals, it emerged that employers expect a variety of new job roles to become more prevalent as 4IR technologies are increasingly adopted across business functions. In the textile and garment manufacturing industry, these include 3D printing specialists, machine learning engineers, and Big Data specialists. In the construction industry, roles in 3D printing and cybersecurity are expected to be created (Figure 31).

Figure 31: Job Roles Expected to Become More Prominent with the Adoption of Industry 4.0 Technologies between 2020 and 2025

Textile and garment manufacturing

3D printing specialists: provide engineering support related to 3D printing and find new ways to use additive manufacturing (e.g., reducing the production time of garment design prototypes)

Machine learning engineers: develop artificial intelligence based algorithms and devices that enable machine learning (e.g., maintain AI-enabled quality control systems)

Big data specialists: set up infrastructure for data collection and analysis, integrate data from various resources to analyze consumer buying trends

Construction

3D printing specialist engineers: provide engineering support and research new applications related to 3D printing (e.g., reducing the production time of mould using 3D printing technology)

Cybersecurity engineers: identify threats and vulnerabilities in systems and software, and develop and implement solutions to reduce the risk of breach and protect confidential building data

AI = artificial intelligence.

Notes: Based on surveys of employers in the textile and garment manufacturing industry (n=51) and information technology-business process outsourcing (IT–BPO) industry (n=51) between June and September 2021; and job data on the IT–BPO industry from the job portal HeadHunter Uzbekistan (accessed 11 June 2021).

Source: Asian Development Bank (Sustainable Development and Climate Change Department).

2 Overview of the Training Landscape

This chapter provides insights into the performance of the technical and vocational education and training (TVET) sector in Uzbekistan as it prepares to deal with the challenges emerging from 4IR technology adoption. The insights are drawn from a survey of training institutions in Uzbekistan, complemented with insights from the employer surveys discussed in Chapter 1.

Overall, training institutions in Uzbekistan have made progress in preparing for 4IR but would require additional support from the government in some areas. Around 47% of training institutions surveyed strongly agree that technical and financial support is needed to enable them to prepare workers for 4IR. In 2020, a significant proportion of training institutions provided courses to improve general digital literacy and over half provided 4IR-specific courses. However, while around 60% of employers in the textile and garment manufacturing and construction industries have adopted IOT technologies, only 30% of training institutions surveyed offer courses in this area. Training institutions would require support to strengthen the industry relevance of their courses specific to 4IR, and leverage 4IR-enabled teaching approaches to improve knowledge delivery. The training landscape analysis also revealed the need for stronger alignment on industry's skills needs between employers and training institutions. While 79% of training institutions assess their graduates to be adequately prepared for entry-level positions, only 47% of employers in the textile and garment manufacturing industry and 65% of employers in the construction industry take the same view. About 54% of training institutions indicated that they would like to see more support for mechanisms for industry collaboration.

Industry 4.0 technologies could lead to rapid changes in the skills that employers require from workers going forward. As such, policies focused on strengthening alignment between employers and training institutions on future skill needs will be critical to ready training institutions to prepare workers for 4IR and ensure that their graduates can gain access to quality jobs.

To better understand the supply of talent and skills for the adoption of 4IR technology, a survey of 70 training institutions was undertaken in Uzbekistan. These include technical colleges and specialized secondary education providers that provide TVET training as well as institutions of higher learning. Of the training institutions surveyed, 98% train at least 100 students per year.

A. Industry 4.0 Readiness and Impact of COVID-19

Close to 70% of training institutions surveyed indicated that they have a good understanding of the skills required for 4IR adoption (Figure 32). However, more than 80% of training institutions said that while they can adequately prepare their worker for 4IR, they would need additional technical and financial assistance to do so. This could be due to training institutions being unclear on the types of 4IR technologies and industry-specific applications prioritized by firms in Uzbekistan, or lacking the financial resources to purchase up-to-date software or equipment to be used to train students in 4IR technologies. During the country consultations conducted, government stakeholders highlighted that the lack of local expertise related to 4IR was a key issue. As the adoption of such technologies was limited among firms in most industries and much of the technology came from outside Uzbekistan, there were few local technology experts who could provide guidance to policy makers and government stakeholders on the design of training curricula. Trainers who were proficient in 4IR technologies were also lacking in Uzbekistan.

Figure 32: Perception of Training Institutions on Readiness for Industry 4.0 in Uzbekistan

Over 80% of training institutions surveyed said they would need additonal technical and financial support to enable them to prepare workers for 4IR

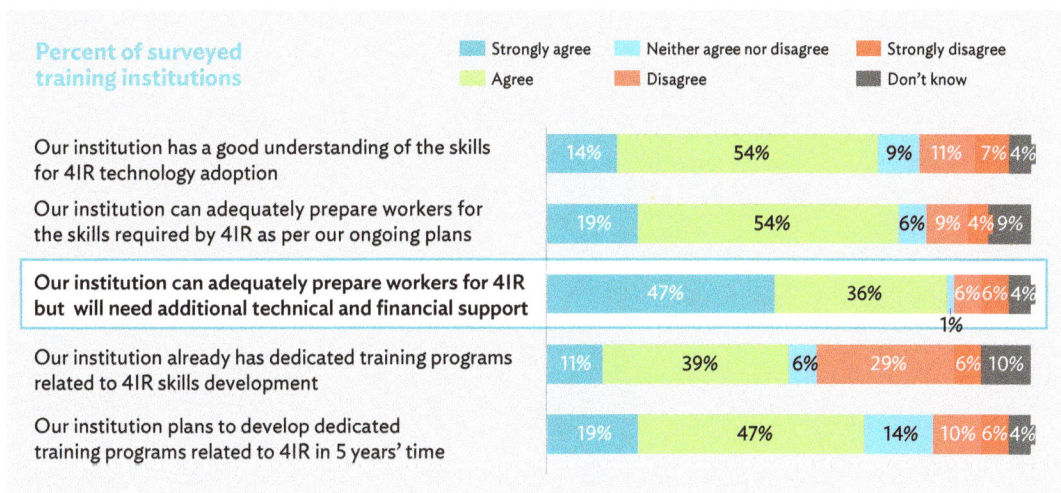

4IR = Fourth Industrial Revolution.
Note: Based on survey of training institutions between June and September 2021 (n=70).
Source: Asian Development Bank (Sustainable Development and Climate Change Department).

Most training institutions in Uzbekistan indicated that they could continue training activities during the COVID-19 pandemic. Only 24% of training institutions surveyed said they had to close fully for some time due to the inability to conduct in-person training and 61% managed to shift some or most of their courses online (Figure 33).

Figure 33: Impact of COVID-19 on Training Institutions in Uzbekistan

Around 60% of training institutions indicated that they shifted some or most of their courses online in response to COVID-19

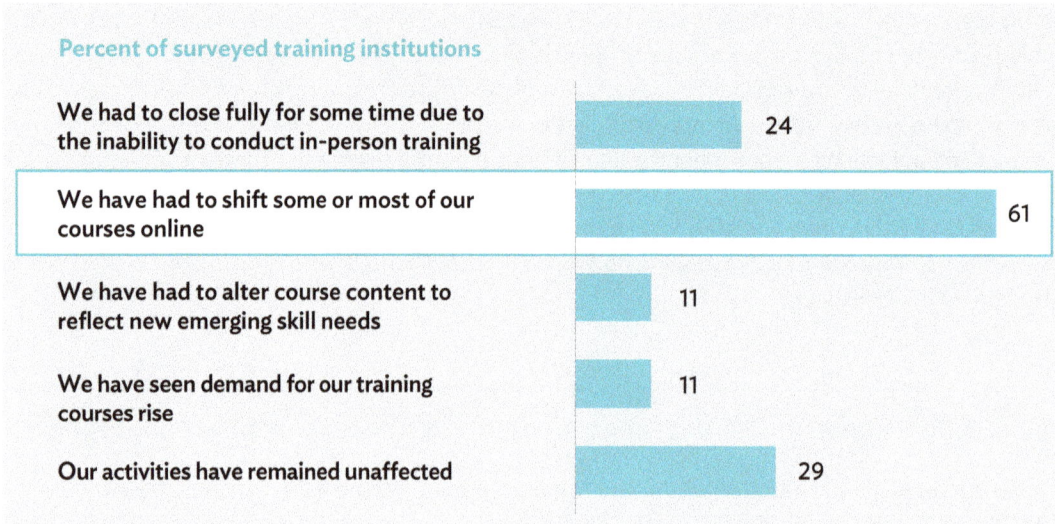

Percent of surveyed training institutions

We had to close fully for some time due to the inability to conduct in-person training	24
We have had to shift some or most of our courses online	61
We have had to alter course content to reflect new emerging skill needs	11
We have seen demand for our training courses rise	11
Our activities have remained unaffected	29

COVID-19 = coronavirus disease.
Note: Based on survey of training institutions between June and September 2021 (n=70). Percentages do not add up to 100% as respondents were asked to select all options that apply.
Source: Asian Development Bank (Sustainable Development and Climate Change Department).

B.   Curricula

Industry 4.0 technologies can transform the workplace and jobs quickly and it is critical that training curricula be updated frequently to meet employers' needs. In Uzbekistan, 97% of training institutions claim that they review and update their training curricula annually or more frequently (Figure 34).

Many training institutions in Uzbekistan claim to already offer 4IR-related courses: 57% offer courses specific to 4IR technologies and 73% provide general digital skills programs to improve digital literacy (Figure 35). Over half of training institutions use online self-learning modules and two-thirds use interactive videos to deliver training. However, only a third use AR or VR mechanisms in their training approaches. Such technologies could make training more effective, particularly as employers shift from OJT to other training channels, as discussed in Chapter 1.

Of the training institutions that provide courses specific to 4IR technologies, most offer courses in Big Data analytics and IOT technologies. However, while around 60% of employers in both industries plan have adopted IOT technologies, only 30% of training institutions offer relevant courses. Another 4IR technology that training institutions could focus on is additive manufacturing. Around a third of employers in the textile and garment manufacturing and construction industries have adopted additive manufacturing technologies, but only 13% of training institutions surveyed offer relevant courses currently. Country consultations with government stakeholders in Uzbekistan point to the need to further strengthen alignment between technology adoption plans of key industries and the design of training curricula to build skills in Uzbekistan's future workforce.

Figure 34: Frequency of Review and Update of Curricula by Training Institutions in Uzbekistan

A majority of training institutions in Uzbekistan update training curricula frequently

Percent of surveyed training institutions

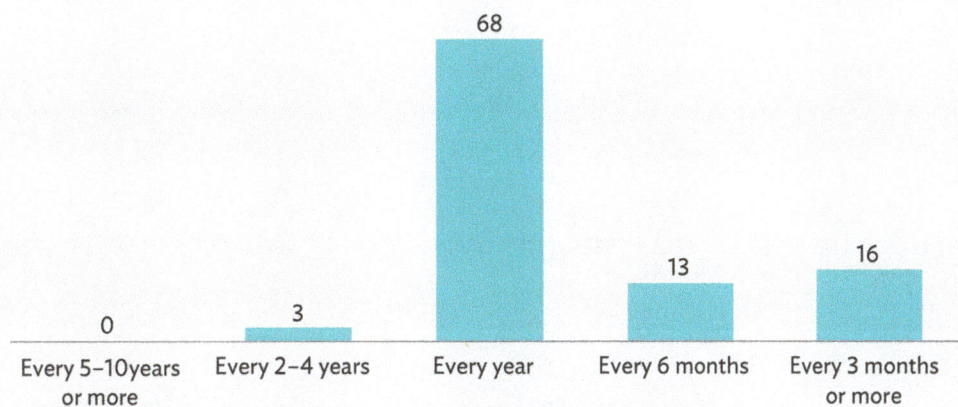

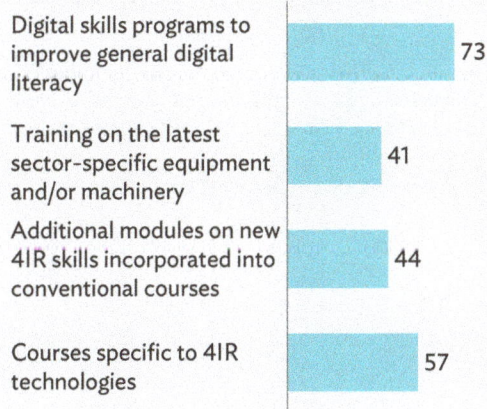

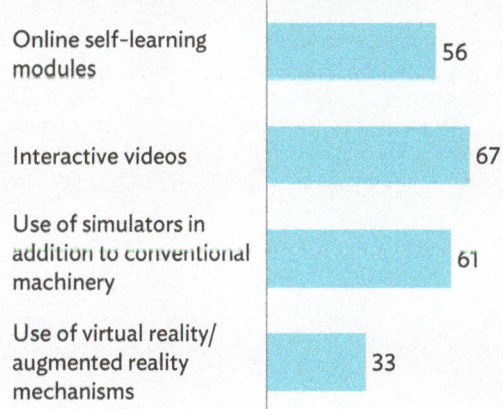

Every 5–10 years or more	Every 2–4 years	Every year	Every 6 months	Every 3 months or more
0	3	68	13	16

Note: Based on survey of training institutions between June and September 2021 (n=70).
Source: Asian Development Bank (Sustainable Development and Climate Change Department).

Figure 35: Prevalence of Industry 4.0-Related Courses and Industry 4.0-Based Delivery in Training Institutions in Uzbekistan

Only 41% of training institutions provide training on the latest sector-specific equipment and/or machinery

Percent of surveyed training institutions

Digital skills programs to improve general digital literacy	73
Training on the latest sector-specific equipment and/or machinery	41
Additional modules on new 4IR skills incorporated into conventional courses	44
Courses specific to 4IR technologies	57

Percent of surveyed training institutions

Online self-learning modules	56
Interactive videos	67
Use of simulators in addition to conventional machinery	61
Use of virtual reality/augmented reality mechanisms	33

4IR = Fourth Industrial Revolution.
Note: Based on survey of training institutions between June and September 2021 (n=70). Percentages do not add up to 100% as respondents were asked to select all options that apply.
Source: Asian Development Bank (Sustainable Development and Climate Change Department).

Sectoral road maps that combine technology adoption and skills development could help to ensure that training institutions can provide students with skills sought after by employers. In Singapore, the Construction Industry Transformation Map sets out the country's vision of an advanced and integrated building sector with widespread adoption of leading technologies that can provide quality jobs. It sets specific objectives of creating quality jobs within the construction industry and enabling workers to transit smoothly into new, higher-skilled roles enabled by technology adoption. Specific policies and programs, including programs to upskill workers, are guided by these objectives across stakeholders in the public and private sector.[9]

Figure 36: Current Adoption of Specific Industry 4.0 Technologies by Employers and Prevalence of Courses Relevant to These Technologies in Training Institutions in Uzbekistan

Around 60% of employers in both industries have adopted IoT technologies but only 30% of training institutions offer relevant courses

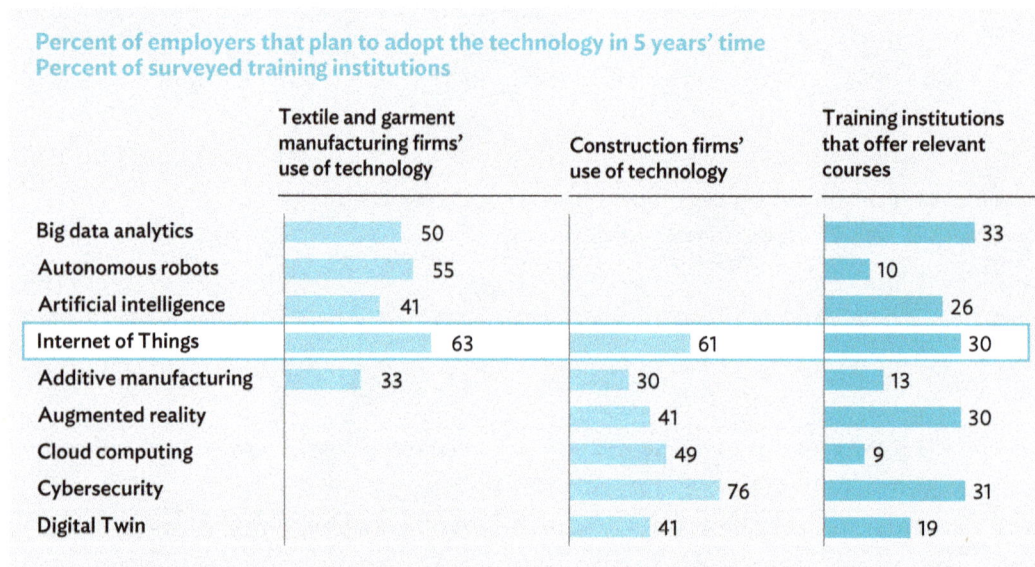

Percent of employers that plan to adopt the technology in 5 years' time
Percent of surveyed training institutions

	Textile and garment manufacturing firms' use of technology	Construction firms' use of technology	Training institutions that offer relevant courses
Big data analytics	50		33
Autonomous robots	55		10
Artificial intelligence	41		26
Internet of Things	63	61	30
Additive manufacturing	33	30	13
Augmented reality		41	30
Cloud computing		49	9
Cybersecurity		76	31
Digital Twin		41	19

AI = artificial intelligence, IT–BPO = information technology–business process outsourcing.

Notes: Based on surveys of employers (n=70 for training institutions; n=51 for textile and garment manufacturing industries; n=51 for construction firms) between June and September 2021. The percentage of employers that have deployed AI technologies is based on the percentage of firms that responded "moderate" or "high" to current deployment of the technology. Percentages do not add up to 100% as respondents were asked to select all options that apply.

Source: Asian Development Bank (Sustainable Development and Climate Change Department).

---

[9]   Ministry of Trade and Industry. ITMs Construction. https://www.mti.gov.sg/ITMs/Built-Environment/Construction.

C.   Industry Engagement

Although training institutions and industry stakeholders in Uzbekistan are not fully aligned on future skills needs created by the adoption of 4IR technologies, collaboration between training institutions and industry stakeholders on current skills needs appear to be strong. Efforts to harness the knowledge of the private sector in training were highlighted during the country consultations conducted with government stakeholders. The European Training Foundation (ETF) cites the sectoral skills councils—which include training institutions and key industry stakeholders—established to develop industry-relevant, professional standards to guide the design of training curricula and improve the employability of graduates (ETF 2021). The Tashkent Construction College (under the Ministry of Construction) is working with TENONICOL Corporation, a provider of building materials and solutions, to establish a training center for construction specialists on modern energy-efficient building materials and construction systems (*UZ Daily* 2021b). Among the training institutions surveyed, 70% indicated that they gather inputs for their curriculum from industry stakeholders, with a large part of such collaboration likely taking place via the sectoral skills councils (Figure 37). A similar proportion of training institutions also work with employers on train-the-trainer programs to foster industry relevance among teaching staff.

Figure 37: Partnership Activities Between Training Institutions and Employers in Uzbekistan

Only 56% of training institutions work with employers to organize workplace-based training for students

Partnership activities between training institutions and employers in Uzbekistan	
Percent of training institutions	**Training institutions**
Gather input for curriculum from industry stakeholders	70
Work with employers on train-the-teacher programs to foster industry relevance	69
Organize workplace-based training for students	56
Organize industry apprenticeships for students	71
Work with employers to place instructors in training/internships with employers to gain practical experience	61
Offer teaching placements for industry professionals at training institutions	61
Work with employers to determine what subjects/disciplines to offer	47
Use employer-provided equipment, facilities, or technology for hands-on training	64
Work with employers to organize job fairs to advertise job opportunities	67

Note: Based on surveys of employers in training institutions (n=70) between June and September 2021. Percentages do not add up to 100% as respondents were asked to select all options that apply.

Source: Asian Development Bank (Sustainable Development and Climate Change Department).

Employer surveys suggest that textile and garment manufacturers work more closely with training institutions compared to employers in the construction industry. Figure 38 shows that 82% of textile and garment manufacturing firms surveyed said they conduct active partnership activities with training institutions compared to 47% of construction firms. While 75% of textile and garment manufacturing firms say they provide input to incorporate the latest industry knowledge in training curricula, only 53% of construction firms did the same. Consistent with the stronger partnership between textile and garment manufacturing firms and training institutions, there is also more communication. More than 70% of textile and garment manufacturing firms communicate with training institutions once a year or more frequently (Figure 39). On the other hand, 31% of construction firms said they never communicate with training institutions.

Figure 38: Partnership Activities Between Employers and Training Institutions in Uzbekistan

A smaller proportion of firms in the construction industry conduct active partnerships with training institutions

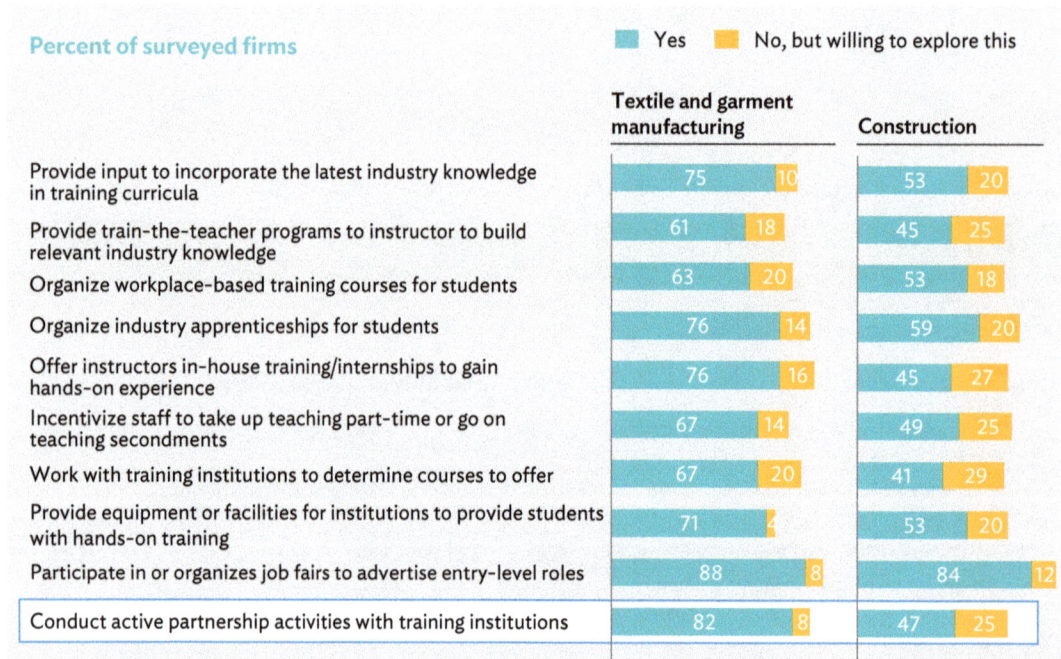

Percent of surveyed firms

Legend: Yes (teal) | No, but willing to explore this (yellow)

	Textile and garment manufacturing (Yes / No)	Construction (Yes / No)
Provide input to incorporate the latest industry knowledge in training curricula	75 / 10	53 / 20
Provide train-the-teacher programs to instructor to build relevant industry knowledge	61 / 18	45 / 25
Organize workplace-based training courses for students	63 / 20	53 / 18
Organize industry apprenticeships for students	76 / 14	59 / 20
Offer instructors in-house training/internships to gain hands-on experience	76 / 16	45 / 27
Incentivize staff to take up teaching part-time or go on teaching secondments	67 / 14	49 / 25
Work with training institutions to determine courses to offer	67 / 20	41 / 29
Provide equipment or facilities for institutions to provide students with hands-on training	71 / 4	53 / 20
Participate in or organizes job fairs to advertise entry-level roles	88 / 8	84 / 12
Conduct active partnership activities with training institutions	82 / 8	47 / 25

Note: Based on surveys of employers (n=51 for textile and garment manufacturing; n=51 for construction) between June and September 2021. Percentages do not add up to 100% as respondents were asked to select all options that apply.

Source: Asian Development Bank (Sustainable Development and Climate Change Department).

Figure 39: Frequency of Communication Between Employers and Training Institutions in Uzbekistan

Thirty-one percent of the employers in the construction industry never communicate with training institutions

Percent of surveyed firms

- Textile and garment manufacturing
- Construction

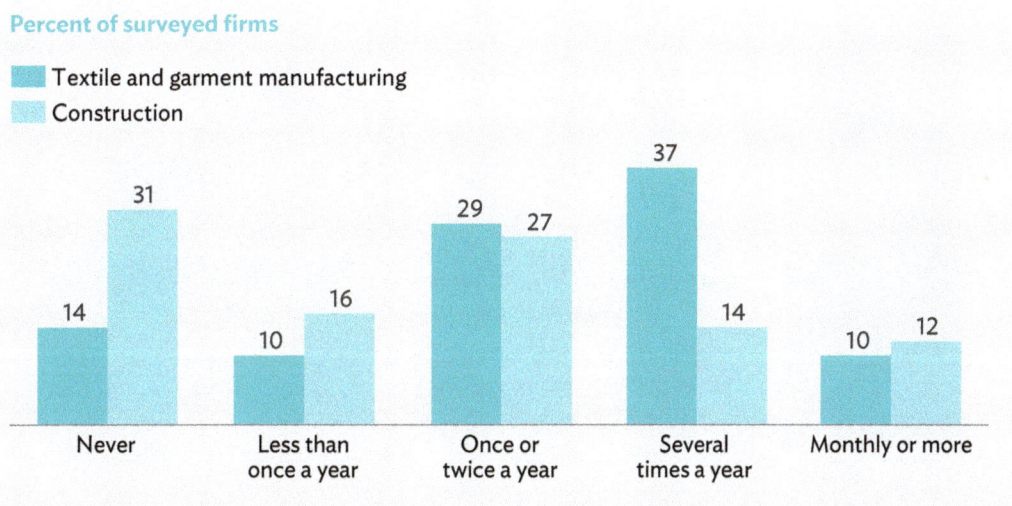

Note: Based on employer surveys (n=51 for textile and garment manufacturing; n=51 for construction) between June and September 2021. Percentages do not add up to 100% as respondents were asked to select all options that apply.
Source: Asian Development Bank (Sustainable Development and Climate Change Department).

D. Teachers, Trainers, and Instructors

The training institution survey revealed some gaps in the efforts of training institutions to ensure the quality of teaching staff. Only 37% of training institutions surveyed conduct reviews of instructors' performance annually or more frequently although 64% conduct frequent feedback sessions with instructors (Figure 40). Institutionalized reviews of instructors' performance can help to ensure that instructors are assessed against common benchmarks and be pegged against monetary incentives such as performance bonuses or salary raises and help improve their performance. Against the backdrop of the adoption of 4IR technology changing jobs and skills needs in the workforce, more professional development initiatives would also be needed to ensure that instructors can take on new training approaches and instruct students in new technologies. Encouragingly, over 80% of training institutions provide professional development and training for their teaching staff currently and these could be leveraged to build 4IR-specific skills or knowledge.

Figure 40: Training Institutions' Practices to Support Instructors in Uzbekistan

Only 37% of training institutions conduct annual/semiannual reviews of instructors' performance

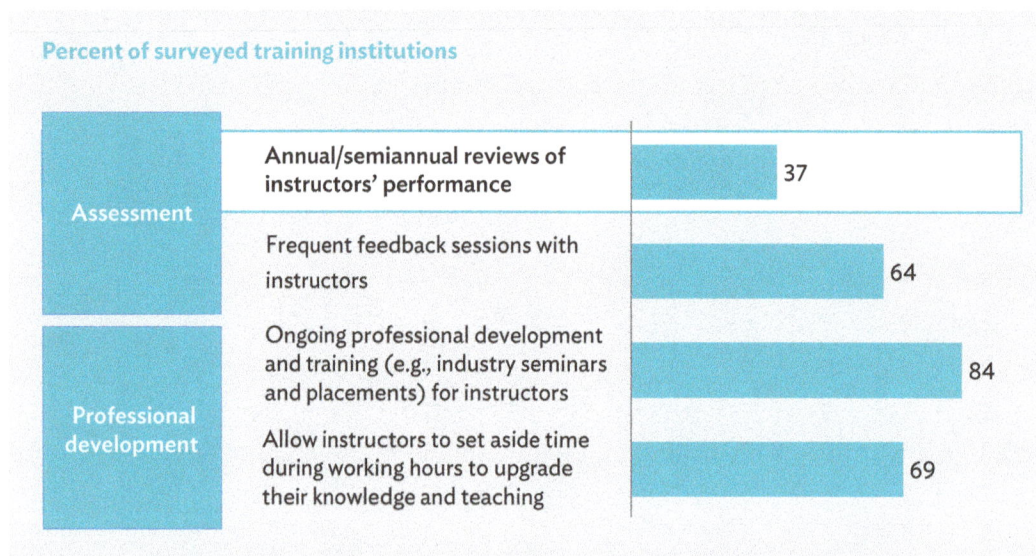

Percent of surveyed training institutions

Assessment	Annual/semiannual reviews of instructors' performance	37
	Frequent feedback sessions with instructors	64
Professional development	Ongoing professional development and training (e.g., industry seminars and placements) for instructors	84
	Allow instructors to set aside time during working hours to upgrade their knowledge and teaching	69

Note: Based on survey of training institutions between June and September 2021 (n=70). Percentages do not add up to 100% as respondents were asked to select all options that apply.
Source: Asian Development Bank (Sustainable Development and Climate Change Department).

E.   Performance and Policy Support

Training institutions were asked to comment on their current performance and the types of policy support that would be required. About 40% of the training institutions said they found it at least somewhat difficult to fill training places, attributing this to students believing that they did not need further training to find jobs and students believing that the institution lacked the capability to help them develop requisite skills (Figure 41). This is consistent with feedback from the employer surveys in Chapter 1 that a large proportion of employers believe that graduates are not workforce-ready, and employers need to provide additional training. In terms of impactful public policies, training institutions cited more government funding for more students to take up courses as well as supportive mechanisms for industry collaboration as useful policy levers (Figure 42).

Figure 41: Training Institutions' Perceptions on and Reasons for Difficulty in Filling Places in Uzbekistan

Forty percent of training institutions find it difficult to fill places—the key reason is students do not think they need more training to find jobs

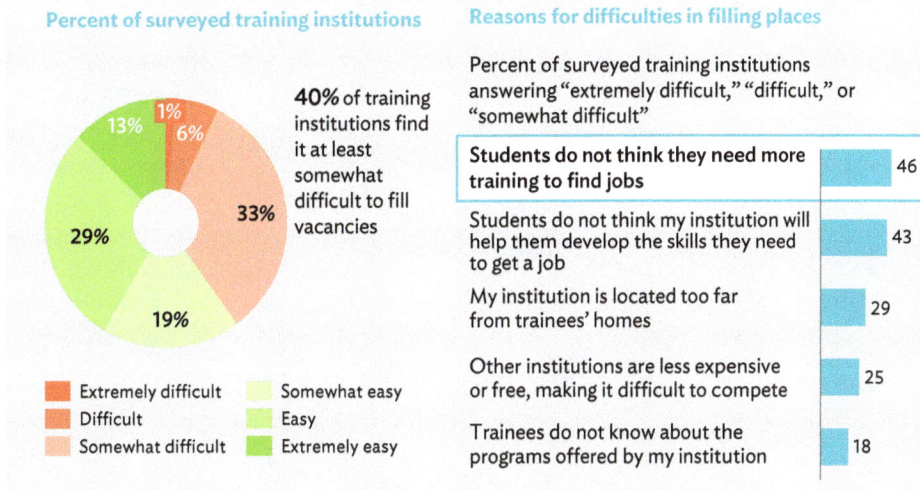

Percent of surveyed training institutions

40% of training institutions find it at least somewhat difficult to fill vacancies

1%
6%
13%
29%
33%
19%

Extremely difficult
Difficult
Somewhat difficult
Somewhat easy
Easy
Extremely easy

Reasons for difficulties in filling places

Percent of surveyed training institutions answering "extremely difficult," "difficult," or "somewhat difficult"

Reason	Percent
Students do not think they need more training to find jobs	46
Students do not think my institution will help them develop the skills they need to get a job	43
My institution is located too far from trainees' homes	29
Other institutions are less expensive or free, making it difficult to compete	25
Trainees do not know about the programs offered by my institution	18

Note: Based on survey of training institutions between June and September 2021 (n=70). Percentages do not add up to 100% as respondents were asked to select all options that apply.

Source: Asian Development Bank (Sustainable Development and Climate Change Department).

Figure 42: Training Institutions' Perceptions on Most Impactful Public Policies for Training Provision in Uzbekistan

Training institutions would like to see more support for students' course fees and mechanisms to strengthen industry collaboration

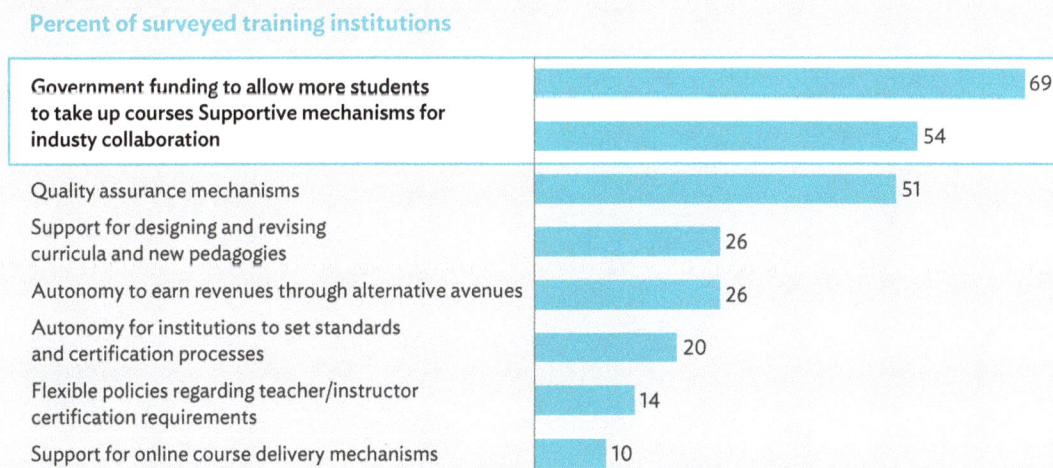

Percent of surveyed training institutions

Policy	Percent
Government funding to allow more students to take up courses Supportive mechanisms for industy collaboration	69
	54
Quality assurance mechanisms	51
Support for designing and revising curricula and new pedagogies	26
Autonomy to earn revenues through alternative avenues	26
Autonomy for institutions to set standards and certification processes	20
Flexible policies regarding teacher/instructor certification requirements	14
Support for online course delivery mechanisms	10

Note: Based on survey of training institutions between June and September 2021 (n=70). Percentages do not add up to 100% as respondents were asked to select all options that apply.

Source: Asian Development Bank (Sustainable Development and Climate Change Department).

F.   Supply and Demand Mismatches

The lack of job openings in the market was cited by training institutions as the most common reason for their graduates being unable to find jobs (Figure 43). Training institutions indicated that they provide substantial support to candidates in their job search: 71% provide support for job applications and 81% organize visits to potential employers (Figure 44). Taken together with the findings on the challenges faced by training institutions in recruiting students, however, it is likely that graduates are unable to find jobs, as employers do not find their qualifications useful. Targeted programs to improve the industry relevance of training curricula could help to address this issue. Encouragingly, Uzbekistan has recently introduced policies to give greater autonomy to training institutions to design and implement training curricula, in alignment with the occupational standards developed by the sectoral skills councils (Government of Uzbekistan 2020f). Previously, training curricula was determined by the government, and training institutions could not update them without undergoing a long approval process of up to 2 years, which often rendered the changes out-of-date by the time they are in effect. The greater autonomy granted to training institutions is expected to substantially improve the responsiveness of training curricula to industry's skill needs.

Figure 43: Training Institutions' Perception of Reasons for Students Being Unable to Find Jobs upon Graduation in Uzbekistan

The lack of information on job openings is a key barrier for graduates finding employment

Ranking based on responses from surveyed training institutions

1 - Most common, 5 – Least common

Rank

1	There are not enough job opportunities
2	Students have insufficient information on job openings in the market
3	Current job opportunities are not attractive enough to incentivize workers to complete relevant training programs
4	Education and training programs do not adequately prepare job seekers for jobs
5	The certifications provided to graduates of training institutions are not well-recognized by employers

Note: Based on survey of training institutions between June and September 2021 (n=70).
Source: Asian Development Bank (Sustainable Development and Climate Change Department).

Figure 44: Non-Training Initiatives Provided by Training Institutions to Support Trainees in Their Professional and Personal Development in Uzbekistan

Seventy-one percent of training institutions in Uzbekistan provide job application and interview support for their students

Percent of surveyed training institutions

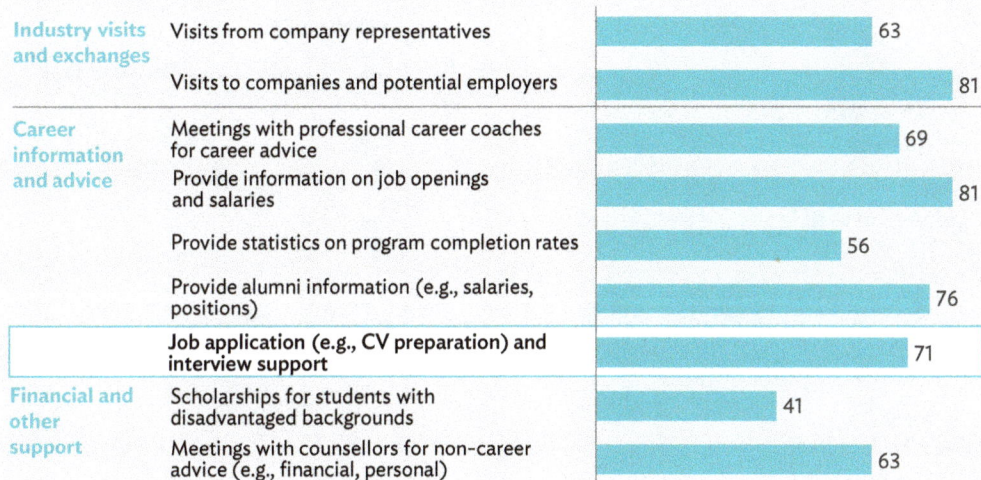

Category	Initiative	Percent
Industry visits and exchanges	Visits from company representatives	63
	Visits to companies and potential employers	81
Career information and advice	Meetings with professional career coaches for career advice	69
	Provide information on job openings and salaries	81
	Provide statistics on program completion rates	56
	Provide alumni information (e.g., salaries, positions)	76
	Job application (e.g., CV preparation) and interview support	71
Financial and other support	Scholarships for students with disadvantaged backgrounds	41
	Meetings with counsellors for non-career advice (e.g., financial, personal)	63

CV = curriculum vitae.

Note: Based on survey of training institutions between June and September 2021 (n=70). Percentages do not add up to 100% as respondents were asked to select all options that apply.

Source: Asian Development Bank (Sustainable Development and Climate Change Department).

There are significant differences between the perceptions of training institutions and employers toward graduate quality in Uzbekistan. Employers in the construction industry have a more positive impression of graduates compared to those in the textile and garment manufacturing industry. While 79% of training institutions surveyed feel that their graduates are adequately prepared for entry-level positions, only 47% of employers in the textile and garment manufacturing industry and 65% of employers in the construction industry agree (Figure 45). Similar differences were observed when training institutions and employers were asked if graduates had the appropriate general and job-specific skills. There could be various reasons for the differences in perception. First, training institutions have an outdated view of industry skill needs and produce graduates that are not skilled in areas relevant to employers. This results in a vast amount of OJT training required to bring graduates up to speed, as found in Chapter 1. The challenge is exacerbated with adoption of 4IR technologies in the textile and garment manufacturing industry rapidly changing skills needs, so that training institutions are unable to keep up with these changes. Second, the lack of regular reviews of instructors' performance means that training institutions are unable to identify gaps in training delivery, including training approaches. It would be critical to address these gaps by ensuring stronger alignment between industry and training institutions on current and future skills needs, and building strong mechanisms for quality assurance in the education system. In addition, 4IR-enabled education technologies could also help to improve the quality of training.

Figure 45: Perception of Employers on Graduates' Preparedness for Entry-Level Positions in Uzbekistan

Training institutions perceive graduates to be better prepared for the workforce than employers, particularly for textile jobs

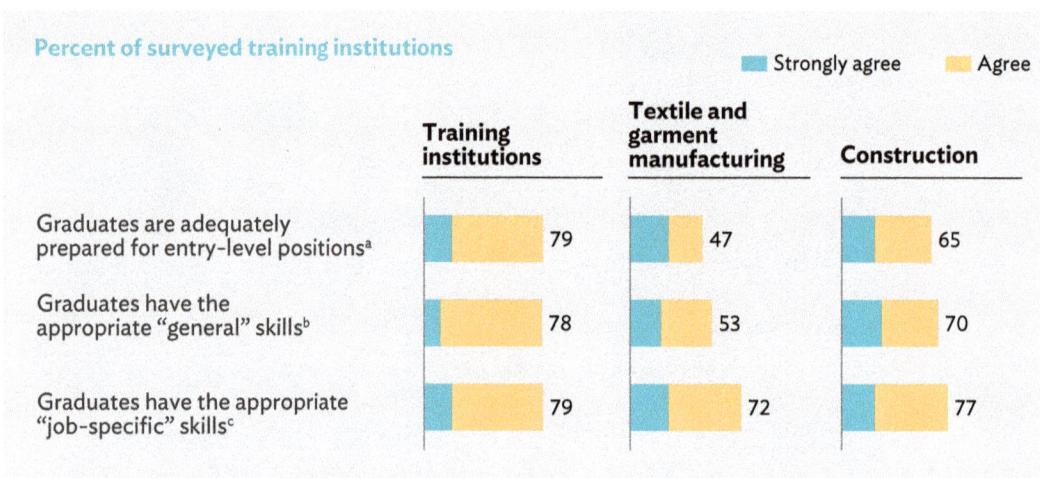

Percent of surveyed training institutions

Strongly agree   Agree

	Training institutions	Textile and garment manufacturing	Construction
Graduates are adequately prepared for entry-level positions[a]	79	47	65
Graduates have the appropriate "general" skills[b]	78	53	70
Graduates have the appropriate "job-specific" skills[c]	79	72	77

[a] In their chosen fields of study.

[b] "General" skills include soft and generic skills that are developed through any academic program and experience and are requisite for success in any job, e.g., teamwork, creativity, and problem solving.

[c] "Job-specific" skills include skills relevant to the discipline or job description that are necessary to succeed in that specific position, e.g., accounting, computer programming, and engineering.

Notes: Based on surveys of employers in textile and garment manufacturing industry (n=51), construction industry (n=51), and training institutions (n=70) between June and September 2021.

Source: Asian Development Bank (Sustainable Development and Climate Change Department).

3 National Policy Responses

A thorough scan of policies and programs across the government, industry, and civil society in Uzbekistan reveals that while a range of strategies have been adopted to foster innovation and increase the digital capabilities of workers, a 4IR road map to guide the adoption of 4IR technologies has not been developed. Uzbekistan's education system has undergone significant reform in recent years and there are a range of efforts to engage employers to create training frameworks that are relevant to industry needs, build effective lifelong learning models, and increase the responsiveness of training curricula to emerging skills needs. Programs have also been implemented to encourage businesses to adopt technology and build digital competencies among firms and workers. However, there is scope for greater coordination between stakeholders, including policy makers, training institutions, and industry stakeholders to ensure that industry development and skills development policies are aligned to enable a smooth transition toward 4IR.

The policy assessment in this chapter leverages a combination of government policy documents, academic literature on Uzbekistan's skills development landscape and relevant government policies, as well as the surveys and skills gap analysis conducted as part of this study.

A. Overview of Industry 4.0 Policy Landscape

The Government of Uzbekistan has adopted a series of strategies and policies that set out the country's vision of building a vibrant knowledge-based economy supported by a skilled workforce with strong digital literacy. The Five-Area Development Strategy, 2017–2021 for Uzbekistan sets out priorities to improve the system of lifelong learning, increase access to quality education, improve workforce employability, and stimulate research and innovation (*The Tashkent Times* 2017). The Digital Uzbekistan 2030 development road map sets out plans to encourage technology adoption across all sectors and improve the digital literacy of the workforce (Government of Uzbekistan 2020a). The Education Sector Plan of Uzbekistan 2019–2023 includes plans to revise education and training curriculum to be more aligned with emerging skills needs and increase the use of ICT in the classroom (Government of Uzbekistan 2019a). In the construction and textile and garment manufacturing industries, a number of decrees and resolutions set out the government's plans to reform these industries, adopt new and innovative technologies, and increase the skill level of workers.[10] Uzbekistan's key 4IR and skills-

---

[10] In the textile and garment manufacturing industry, these include a presidential decree to support the textile, sewing, and knitting industry; a presidential decree for the further development of light industry and stimulation of the production of finished products; a resolution of the Cabinet of Ministers for the further expansion of mechanisms for the introduction of cotton and textile production in the republic; and a resolution of the Cabinet of Ministers for the further development of cotton and textile production. In the construction industry, these include a resolution of the Cabinet of Ministers to reform customer service in the field of capital construction, and a resolution of the Cabinet of Ministers on evaluation of seismic strength of buildings and structures and the introduction of the system for formation of electronic technical passports.

related government policies, including industry-specific policies for the textile and garment manufacturing and construction industries, include the following:

(i) **Five-Area Development Strategy, 2017–2021.** This is a development road map for Uzbekistan that sets out its priorities in public administration reform, economic growth, as well as socioeconomic development. Specific to education, the strategy sets out plans to improve the system of lifelong learning, increase access to quality education, improve workforce employability, and stimulate research and innovation (*The Tashkent Times* 2017).

(ii) **State Program for Year of Youth Support and Health Promotion in 2021.** The development plan sets out the specific measures that Uzbekistan will undertake in 2021 to meet the objectives of the Five-Area Development Strategy, 2017–2021. It focuses on plans to strengthen the support for youth and create the foundation for long-term economic growth. It sets out initiatives to foster entrepreneurship and improve the quality and effectiveness of lifelong education among other areas (*Review.Uz* 2021).

(iii) **Digital Uzbekistan 2030.** This is a digital development road map for Uzbekistan that sets out the government's plans to develop the digital economy and encourage technology adoption across all sectors. It includes plans for measures to develop the digital industry and improve the digital literacy of the general population (Government of Uzbekistan 2020a).

(iv) **Uzbekistan 2035.** This is a development strategy framework for Uzbekistan to 2035 formulated by the Buyuk Kelajak.[11] It lays out strategies for Uzbekistan to adopt technology in various sectors as well as measures for the development of human capital. Specific initiatives include promoting the use of 4IR technologies such as artificial intelligence, automation, and robotics technologies in various industries to increase productivity, as well as the establishment of effective lifelong learning models to constantly upskill the workforce (UzDaily 2019b).

(v) **Strategy for Innovative Development for 2019 to 2021.** The strategy is designed to improve research excellence, strengthen the links between education, science, and industry, and increase public and private investments in innovation, research and development, and modern technologies (Government of Uzbekistan 2018).

(vi) **Concept of Development of the Higher Education System in Uzbekistan until 2030.** This is a development road map setting out plans to align higher education curriculums more closely with industry needs and to improve access to higher education. It also sets out plans for the establishment of the Republican Council of Higher Education tasked to analyze international best practices and develop proposals and measures to improve the quality of higher education in Uzbekistan (Government of Uzbekistan 2019b).

(vii) **Education Sector Plan of Uzbekistan 2019–2023.** This is a development strategy for Uzbekistan's education system from 2019 to 2023, and sets out key strategic objectives including the revision of training curriculum to be more aligned with skills needs, improvement of professional support and capacity building of teachers, and the increased use of ICT in the classroom (Government of Uzbekistan 2019a).

(viii) **Resolution of the President on Measures for the Cardinal Enhancement of the System of Assessment of Qualifications and Providing the Labor Market with Qualified Personnel.** This policy document sets out measures to improve the quality and relevance of higher and professional education and create a skilled workforce able to meet industry needs. It lays out the establishment of

---

[11] Buyuk Kelajak (great future) is international nongovernmental nonprofit organization, established by Uzbekistan compatriots and patriots living abroad, that aims to introduce innovative ideas into the industry and science of Uzbekistan. The organization plans to develop a long-term strategy for Uzbekistan until 2035 in the banking, industrial, financial, educational, investment, and technological areas.

sectoral skills councils for the development of professional qualifications and competencies, and tasks the industry councils to develop and implement professional standards relevant to industry needs (Government of Uzbekistan 2020c).

(ix) **Law about Employment of the Population.** This is a policy document that sets out the direction of state policy in employment. It lays out plans to provide support for jobseekers and unemployed workers, particularly vulnerable groups, including through professional training and reskilling (Government of Uzbekistan 2020d).

(x) **Strategy of Modernization and Innovative Development of the Construction Industry for 2021–2025.** The strategy sets out plans to transform the construction industry through modern management methods, the adoption of innovative technologies, and human capital development. (Government of Uzbekistan 2020b).

(xi) **Presidential Decree About Measures for the Accelerated Development of the Textile, Sewing, and Knitting Industry.** The decree sets out a road map for the development of textile and garment manufacturing industry, including plans to strengthen the workforce by identifying in-demand skills in the industry and increasing practical, workplace-based training. It also forms a working commission to develop further plans for the development of the cotton and textile industries (UZ Daily 2017).

(xii) **Presidential Decree About Measures for the Urgent Support of the Textile, Sewing and Knitting Industry.** The decree sets out plans to counter the impact of COVID-19 on the textile and garment manufacturing industry, including through involving international experts to drive reforms in the industry to increase productivity and programs to attract foreign investments (Government of Uzbekistan 2020e).

Key policies relevant to managing the impact of 4IR on skills in Uzbekistan are summarized in Table 6.

Table 6: Key Policies Relevant to Managing the Impact of Industry 4.0 on Skills in Uzbekistan

Policy Document	Responsible Entity/Entities	Relevance
Five-Area Development Strategy, 2017–2021	A National Commission headed by President Shavkat Mirziyoev is responsible for implementation of the strategy	A development road map for Uzbekistan that sets out plans to improve the system of lifelong learning, increase access to quality education, improve workforce employability, and stimulate research and innovation
State Program for Year of Youth Support and Health Promotion in 2021	Prime Minister and Cabinet of Ministers	A development plan for 2021, focusing on plans to strengthen support for youth and create the foundation for long-term economic growth. It sets out initiatives to foster entrepreneurship and improve the quality and effectiveness of lifelong education among other areas.
Digital Uzbekistan 2030	Ministry for Development of Information Technologies and Communications	A digital development road map for Uzbekistan that includes plans for measures to develop the digital industry and improve digital literacy for the general population
Uzbekistan 2035	Buyuk Kelajak	A development strategy framework that lays out strategies for Uzbekistan to adopt technology in various sectors as well as measures for the development of human capital

continued on next page

Table 6 *continued*

Policy Document	Responsible Entity/Entities	Relevance
Strategy for Innovative Development, 2019–2021	Ministry of Innovation Development	The strategy is designed to improve research excellence, strengthen the links between education, science, and industry, and increase public and private investments in innovation, research and development, and modern technologies
Concept of Development of the Higher Education System in Uzbekistan until 2030	Ministry of Federal Education and Professional Training	A development road map that sets out plans to establish the Republican Council of Higher Education that will analyze international best practices and develop proposals and measures to improve the quality of higher education in Uzbekistan
Education Sector Plan of Uzbekistan, 2019–2023	Ministry of Public Education	A development strategy for Uzbekistan's education system from 2019 to 2023 that sets out key strategic objectives including the revision of training curriculum to be more aligned with skills needs and the increased use of ICT in the classroom
Resolution of the President on Measures for the Cardinal Enhancement of the System of Assessment of Qualifications and Providing the Labor Market with Qualified Personnel	The Prime Minister and Cabinet of Ministers	Policy document setting out measures to improve the quality and relevance of higher and professional education and create a skilled workforce able to meet industry needs
Law about Employment of the Population	Ministry of Employment and Labor Relations	Policy document that sets out the direction of state policy in employment. It lays out plans to provide support for jobseekers and unemployed workers, particularly vulnerable groups, including through professional training and reskilling

ICT = information and communication technology.

Sources: *The Tashkent Times*. 2017. Uzbekistan's Development Strategy for 2017–2021 Has Been Adopted Following Public Consultation. 8 February. http://tashkenttimes.uz/national/541-uzbekistan-s-development-strategy-for-2017-2021-has-been-adopted-following-discussion; Review.Uz. 2021. State Program "Year of Support for Youth and Health Promotion" (Государственная программа «Год поддержки молодёжи и укрепления здоровья населения»). 3 February. https://review.uz/post/gosudarstvennaya-programma-god-podderjki-molodyoji-i-ukrepleniya-zdorovya-naseleniya; Government of Uzbekistan. 2020a. *Presidential Decree on Approval of the Strategy "Digital Uzbekistan 2030" and Measures for its Effective Implementation* (translated from the Uzbek language). Tashkent. https://lex.uz/docs/5030957#-5031880 https://uzdon.uz/uz/news/info/uzbekistan/165/; Buyuk Kelajak. 2019. Uzbekistan 2035. Tashkent. https://uzbekistan2035.uz/uzbekistan-2035/; Government of Uzbekistan. 2019b. *Presidential Decree on Approval of the Concept of Development of the Higher Education System of Uzbekistan until 2030* (translated from the Uzbek language). Tashkent. On Approval of the Concept of Development of the Higher Education System of Uzbekistan until 2030—The Decree of the President of the Republic of Uzbekistan (translated from the Uzbek language). http://www.erasmusplus.uz/images/shared/file/Concept%20of%20Higher%20Education%20of%20Uzbekistan%20until%202030_EN%20_NEO%20UZ.pdf; Government of Uzbekistan. 2019a. *Education Sector Plan of Uzbekistan 2019–2023* (translated from the Uzbek language). Tashkent. https://www.globalpartnership.org/content/education-sector-plan-2019-2023-uzbekistan; Government of Uzbekistan. 2020c. *Resolution of the President on Measures for the Cardinal Enhancement of the System of Assessment of Qualifications and Providing the Labor Market with Qualified Personnel* (translated from the Uzbek language). Tashkent. https://cis-legislation.com/document.fwx?rgn=129993; and Government of Uzbekistan. 2020d. *Law about Employment of the Population* (translated from the Uzbek language). Tashkent. https://cis-legislation.com/document.fwx?rgn=128048.

B.   Assessment of Current Policy Approaches in Uzbekistan Related to Industry 4.0 and Skills

A diagnostic approach was taken to understand two important aspects of the 4IR skills policy approach in Uzbekistan: (1) "the what": specific policies being adopted by Uzbekistan and how they compare to international best practice approaches in preparing workers for 4IR; and (2) "the how": the implementation mechanisms supporting 4IR efforts in government.

Assessment of Policy Actions ("The What")

Policies and programs in Uzbekistan have been grouped into three action agenda that are assessed to be most crucial to managing the impact of 4IR on jobs and skills.[12] Figure 46 shows the current degree of focus by the country for each action area. The current degree of focus on each action area has been rated as "strong,"

Figure 46: Degree of Focus of Policy Actions to Manage the Impact of Industry 4.0 on Jobs and Skills in Uzbekistan

Degree of focus of policy actions to manage the impact of 4IR on jobs and skills in Uzbekistan

Degree of current focus:[a] ■ Strong   ■ Moderate   ■ Weak

Action agenda	Key action	Assessment
Stimulate 4IR adoption and worker reskilling efforts	Ensure strong and even adoption of 4IR across firms and workers	Strong
	Build awareness of "in-demand" jobs and skills, as well as the benefits and opportunities of training	Moderate
	Incentivize employers and workers to participate in skills development	Strong
	Foster close collaboration between governments, industry, and civil society to create relevant and effective nation-wide training frameworks	Moderate
Create new flexible qualification pathways	Establish effective lifelong learning models	Moderate
	Ensure relevance and agility of education and training curricula to emerging skill needs	Moderate
	Encourage focus on skills rather than just qualifications in both recruitment and national labor market strategies	Moderate
Build inclusiveness to extend 4IR benefits to all workers	Build inclusive models that allow underserved groups to benefit from 4IR	Moderate
	Create social protection mechanisms for workers taking on flexible forms of labor	Weak

4IR = Fourth Industrial Revolution.

Notes: Degree of focus was assessed based on the following criteria: "Strong": few or no gaps between the country's coverage of policy actions and coverage seen in international best practices; "Moderate": medium level of gaps between the country's coverage of policy actions and coverage seen in international best practices; and "Weak": significant gaps between the country's coverage of policy actions and coverage seen in international best practices.

Sources: Authors' representation; AlphaBeta analysis.

---

[12] Based on AlphaBeta research on international best practices for policy actions that manage the impact of Industry 4.0 on jobs and skills. For details of these best practices, see Microsoft and AlphaBeta. 2019. *Preparing for AI: The Implications of Artificial Intelligence for Jobs and Skills in Asian Economies.* https://news.microsoft.com/apac/2019/08/26/preparing-for-ai-the-implications-of-artificial-intelligence-for-jobs-and-skills-in-asian-economies/.

"moderate," or "weak" based on the analyzed extent of the policies' coverage in terms of scope and scale, as compared to those observed in international best practices.

Overall, the current degree of focus varies across the 4IR-relevant policy areas in Uzbekistan. A range of strategies have been adopted in Uzbekistan to foster innovation and increase the digital capabilities of workers. There are also efforts to engage employers to create training frameworks and curricula that are relevant to industry needs. However, there is scope for policies to better incorporate 4IR trends and ensure that businesses and workers in Uzbekistan can benefit from the adoption of 4IR technologies.

(i)   **Stimulating 4IR adoption and worker reskilling efforts.** Various programs and government decrees in Uzbekistan set out plans to build the digital literacy of the workforce and modernize industry through the adoption of advanced technologies and innovation. The Digital Uzbekistan 2030 strategy will subsidize the costs for Uzbeks to obtain educational qualifications in IT-related fields (Government of Uzbekistan 2020a). The One Million Uzbek Coders project provides free distance learning courses to train domestic programmers and the Future Skills Uzbekistan program aims to increase the number of qualified IT specialists through a comprehensive training program covering areas such as mobile application development, IT security and network technologies, and 3D modeling (UzDaily 2019a). To foster innovation in the construction industry, the Tashkent Institute of Architecture and Construction has constructed digital laboratories, and consultations with government stakeholders reveal a strong focus on creating an enabling environment for domestic innovation going forward (Government of Uzbekistan 2020g). Efforts to support firms to adopt technologies and train workers were also highlighted at the country consultations conducted. Financial incentives are provided to enterprises and startups to provide internships and jobs to youths and train newly hired workers (Government of Uzbekistan 2020a). To foster closer collaboration between government and industry on skills development, sectoral skill councils tasked to develop professional standards to guide training curricula have been established in various sectors. This includes the council for professional qualifications in the textile, sewing, and leather industries (ETF 2021). To further strengthen existing efforts, policy makers could consider implementing robust mechanisms to improve awareness of current in-demand skills and anticipate potential skill gaps. Research shows that the matching of jobs with skills is weak in Uzbekistan as labor market information systems (e.g., job portals) are limited and not regularly updated (ADB 2020b). Consultations with government stakeholders further revealed that real-time or updated data on changing skill needs is not available, and it would be important to conduct regular skill needs surveys or put in place mechanisms to collect such data. The establishment of the National Institute for Labor Market Research to collect and analyze real-time data on the labor market in close collaboration with industry stakeholders marks a positive step in addressing the gap and policy makers could ensure that the mechanisms implemented by the Institute consider the rapid skill changes that will accompany 4IR adoption, lending on the support of development partners where required.

(ii)   **Creating new flexible qualification pathways.** There are ongoing efforts in Uzbekistan to develop effective lifelong learning models and focus the education system on building industry-relevant skills. For instance, the National Qualifications Framework defines the knowledge and skills necessary at each professional level, which can be achieved through a variety of pathways, including formal learning as well as work experience (ETF 2021). Government agencies have been tasked to work closely with industry-specific councils to develop professional standards focused on practical skills and revise the training curricula of institutions in alignment with these standards (ETF 2021). Efforts have also been taken to ensure the relevance of training curricula to emerging skills needs and a series of policy reforms have been undertaken to provide greater autonomy to training institutions to update and shape training curricula, including with the support of overseas partners and talent from abroad (Government of Uzbekistan 2020h). There are also plans to improve the digital literacy of the population, particularly youth and women, with the Digital Uzbekistan 2030 strategy setting out plans to establish digital technology training centers across Uzbekistan (Government of Uzbekistan 2020a). There are some areas in which policy focus could be strengthened to ensure that workers are able to meet the changing skill demands created by 4IR. First, employer surveys show that textile and garment manufacturing and construction firms face challenges in hiring workers who are adequately prepared for their jobs and need to invest substantially in OJT for their workers. Training institutions could partner industry stakeholders more closely to ensure that training is relevant. Second, training infrastructure for continual reskilling of adult learners could be strengthened. Past work shows that continuing education in Uzbekistan is largely associated with only formal education, mainly for people to upgrade their qualifications (ADB 2020b). With the rapid skills demand changes created by 4IR, short-term courses and micro-learning provide alternatives for the workforce to gain skills quickly and policy makers could consider policies to enable the recognition of such qualifications and make relevant courses widely available.

(iii) **Building inclusiveness to extend 4IR benefits to all workers.** Uzbekistan has strong social protection networks to support vulnerable groups and unemployed workers. The government maintains records of families, youths, and women in need of financial assistance and support.[13] Vulnerable youths and women are provided with course fee subsidies to attend training courses at private vocational training centers (Market Screener 2020). In addition, there are a range of initiatives to provide training and job placement for unemployed workers. In 2019, the Ishga Marhamat center was created in Tashkent, as a pilot project to provide job advisory and training services to unemployed persons (ETF 2021). Existing social protection policies could be strengthened by considering the changes in skills needs created by 4IR. For instance, seamstresses whose sewing skills are made obsolete by automated sewing machines could be trained to operate such machines and conduct quality checks. In addition, there is currently limited focus on robust social protection mechanisms for digital freelancers and gig economy workers, as well as other new, flexible forms of labor that emerge with 4IR adoption.

Assessment of Implementation of 4IR Policies ("The How")

The implementation of a 4IR strategy for jobs and skills in Uzbekistan was assessed against three dimensions found to be crucial for success according to past academic work: the clarity and robustness of plans, the strength of coordination between different stakeholders, and the alignment of financing and incentives (Figure 47).[14]

Overall, the implementation approach varies across the critical dimensions. Uzbekistan has established clear strategies for the development of an innovation-driven economy and developed plans to encourage the adoption of 4IR technologies in the textile and garment manufacturing and construction industries. Efforts have also been undertaken to improve coordination between government departments, training institutions, and industry in skills development, particularly through the establishment of sectoral skills councils. There is scope for Uzbekistan to build on these efforts and strengthen linkages between industry development plans and skills development plans. More specifically:

(i) **Clarity and robustness of plans.** The Five-Area Development Strategy for 2017–2021 and Digital Uzbekistan 2030 road map sets out broad plans for technology adoption across various sectors and

---

[13] Uzbekistan maintains a list of families in need of financial assistance and support, also known as the "iron notebook." Similarly, a "youth notebook" and a "women's notebook" is maintained in each district, city, and region. The notebooks include young people and women in need of social, legal, and psychological support, as well as those seeking to acquire new knowledge and professions.

[14] Based on AlphaBeta research of Industry 4.0 strategies, plus insights from past public sector research, including: Barber. 2007. https://books.google.com.sg/books/about/Instruction_to_Deliver.html?id=MLcbAQAAMAAJ&redir_esc=y. *Instruction to Deliver: Fighting to Transform Britain's Public Services; McKinsey & Company. 2012. Delivery 2.0: The New Challenge for Governments.* https://www.mckinsey.com/industries/public-sector/our-insights/delivery-20-the-new-challenge-for-governments.

Figure 47: Implementation Challenges Associated with Industry 4.0 Policies for Jobs and Skills in Uzbekistan

Implementation challenges associated with 4IR policies for jobs and skills in Uzbekistan

Degree of current focus:[a] ■ Strong ■ Moderate ■ Weak

Dimension	Questions	Assessment
Clarity and robustness of plans	Is there a clearly articulated vision for 4IR?	Weak
	Is there strong integration between employment/skills and the 4IR plan?	Weak
	Is the plan forward looking, incorporating 4IR trends?	Weak
	Is there strong local data to support evidence-based policymaking?	Moderate
Strength of coordination	Is there one shared road map across industry and government departments for 4IR?	Weak
	Is there coordination across different government ministries and levels?	Moderate
	Is there strong alignment within and between industry, and education and training institutions?	Moderate
Alignment of financing & incentives	Is government financing aligned with the strategic goals?	Strong
	What are the strength of incentives for employers and workers to invest in skill development? What are the strength of incentives for teachers and institutions to ensure high-quality training and education systems?	Strong

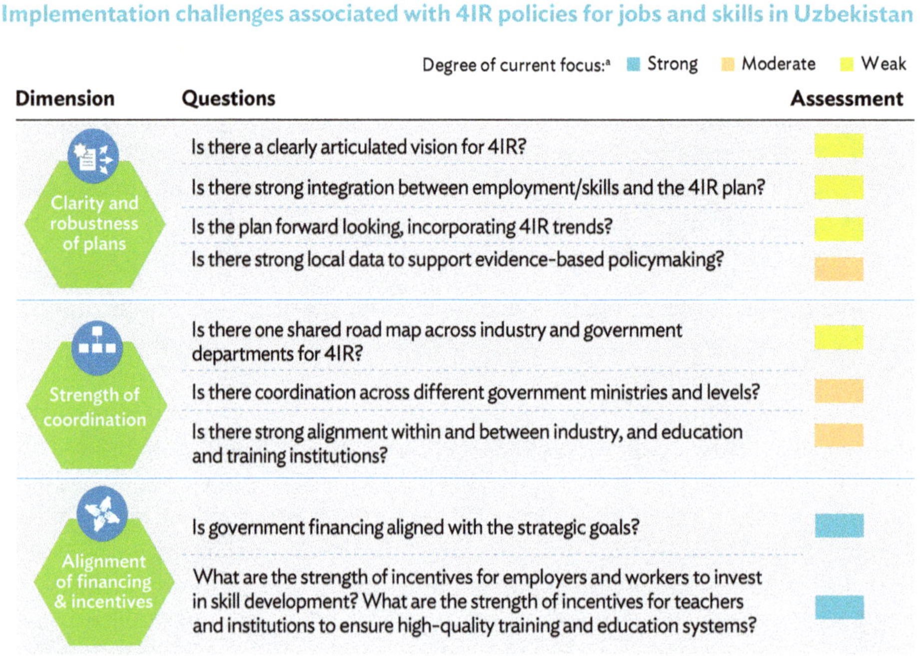

4IR = Fourth Industrial Revolution.

Notes: Degree of focus was assessed based on the following criteria: "Strong": few or no gaps between the country's policy implementation approach and approach seen in international best practices; "Moderate": medium level of gaps between the country's policy implementation approach and approach seen in international best practices; and "Weak": significant gaps between the country's policy implementation approach and approach seen in international best practices.

Sources: Authors' representation; AlphaBeta analysis.

strategies to improve digital literacy among the population (Government of Uzbekistan 2020a). In 2017, the Ministry of Innovation Development was established to drive the development of an innovation-driven economy and the Strategy for Innovative Development for 2019 to 2021 was subsequently adopted, with the objective of strengthening the links between education, science, and industry, and increasing public and private investments in innovation and modern technologies. Decrees focused on the textile and garment manufacturing and construction industries further set out specific policies to encourage the adoption of 4IR technologies in each industry. Amid various government strategies and decrees aimed at helping firms in Uzbekistan transform digitally, there are two ways in which the clarity and robustness of Uzbekistan's 4IR plans could be strengthened. First, there is no single action plan setting out the long-term growth vision for 4IR in each industry that consolidates plans for innovation, technology adoption, and skills development. Existing plans and policies consider these aspects of growth separately and might not be able to adequately address the skills development needs created by the transition toward 4IR. Second, access to strong local data to support evidence-based policymaking must be a priority. While the establishment of the National Institute for Labor Market Research marks a positive step in the collection and analysis of real-time data related to skills needs and gaps in the labor market, further efforts could be taken to ensure that data collected will also be relevant to the design of 4IR-related jobs and skills as highlighted at the country consultations conducted with government stakeholders.

(ii) **Strength of coordination.** Efforts have been taken to improve coordination on skills development and job creation among government agencies, training institutions, and industry stakeholders. Uzbekistan recently reformed its education system to strengthen linkages between training and jobs. Under these reforms, each ministry was tasked to establish training frameworks and programs for the industry under its purview, in cooperation with the newly established sectoral skills councils and training institutions (i.e., universities and vocational training centers) (ETF 2021). However, there is scope to improve the coordination among stakeholders for 4IR adoption. Plans for skills development could be more clearly aligned with strategies to drive innovation and technology adoption in each industry and a clear lead agency identified to coordinate these efforts in each industry.

(iii) **Alignment of financing and incentives.** Government funding for education in Uzbekistan was 5.9% of gross domestic product in 2018, as compared to the 4.7% average for Central and South Asian countries (ADB 2020a). In recent years, Uzbekistan has introduced significant education reforms and strengthened incentives for training institutions to improve the quality of education and training. Prior to education reforms in 2019, vocational education formed part of the 12 years of compulsory education in Uzbekistan and was fully state funded with admission quotas established centrally. Since 2019, compulsory education has been reduced to 11 years and vocational education made optional, so that institutions would need to compete for students, incentivizing them to improve the quality of training programs (ETF 2021).

C.   Assessment of 4IR Policies in Relation to the COVID-19 Pandemic

Various strategies were adopted in Uzbekistan to foster innovation and improve the digital literacy of businesses and workers during the pandemic. Around 50% of firms surveyed expect the COVID-19 to accelerate the adoption of 4IR technologies across the textile and garment manufacturing and construction industries in Uzbekistan. As demonstrated in Chapter 1, the adoption of 4IR could change task profiles and skill needs in both industries significantly.

One key strategy to build digital literacy in the workforce is the Future Skills Uzbekistan program (UZ Daily 2021d). The program aims to increase the number of qualified IT specialists through a comprehensive training program covering areas such as mobile application development, IT security and network technologies and 3D modeling among other areas. Subsidies are provided for unemployed citizens to participate in the program through the Ministry of Employment and Labour Relations Fund for Assistance to Employment of the Population. To ensure that businesses can respond effectively to the pandemic, including via adopting digital technologies, the Uzbekistan Chamber of Commerce and Industry worked with the United Nations Development Programme to launch a Business Clinic program targeted at SMEs and entrepreneurs (UzDaily 2020). The Business Clinic operates in the form of a telegram channel, a phone hotline, and a website, and provides 24-hour consultation services to businesses on business management and government support schemes. The government also rolled out the Digital Uzbekistan 2030 road map that reiterated its commitment toward improving technology adoption across sectors and the population's digital literacy, showing a long-term commitment to postpandemic digital transformation (Government of Uzbekistan2020a).

4 The Way Forward

The previous three chapters highlight the potential benefits that 4IR can bring to Uzbekistan as well as the challenges that would need to be addressed to achieve these benefits. This chapter identifies policy recommendations, based on global best practices, that policy makers in Uzbekistan can consider adopting, to unleash the potential opportunities created by 4IR.

A. Summary of Key Challenges to Industry 4.0 Adoption Faced by Uzbekistan

Table 7 provides a recap of the challenges facing Uzbekistan from the industry analysis (Chapter 1), the training institution survey (Chapter 2), and the policy assessment (Chapter 3).

Table 7: Recap of Challenges Facing Uzbekistan in Relation to Industry 4.0

Area	Key Challenges	Findings
Textile and garment manufacturing industry	The adoption of 4IR will rapidly change task profiles and skills needs in the textile and garment manufacturing industry	Creative thinking and/or design and digital and/or ICT skills will become the most valued skills by employers by 2025, with 4IR adoption
	Training institutions do not offer courses in 4IR technologies relevant to textile and garment manufacturing employers	Although **55%** of textile and garment manufacturing firms have adopted autonomous robots, only **10%** of training institutions offer relevant courses
	Firms face challenges in hiring graduates that are adequately prepared for their roles	Only **47%** of textile and garment manufacturers agree that graduates hired in the past year were adequately prepared for the job

continued on next page

Table 7 *continued*

Area	Key Challenges	Findings
Construction Industry	Understanding of 4IR technologies varies significantly across firms	**35%** of construction firms have an advanced understanding of 4IR technologies, but a similar proportion have not heard of 4IR
	Limited use of VR/AR technologies in the classroom despite the potential for such technologies for construction-related training	Only **33%** of training institutions use VR/AR technologies to deliver training
	Job gains from 4IR benefit male workers significantly more than female workers	The number of jobs expected to be gained by male workers exceeds the number of jobs expected to be gained by female workers by **6.5x**
Training institutions	Training institutions require additional resources to prepare workers for 4IR	**47%** of training institutions strongly agree that additional technical and financial support is needed to enable them to prepare workers for 4IR
	Few training institutions provide relevant 4IR-specific courses on 4IR	Close to **90%** of firms plan to adopt IOT technologies but only **30%** of training institutions offer relevant courses
	Limited use of 4IR technologies	Only **a third** of training institutions use augmented reality or virtual reality mechanisms to deliver training
Policy assessment	Lack of clearly articulated 4IR vision integrated with jobs and skills	Plans to improve education and vocational training systems do not incorporate 4IR trends
	Lack of targeted programs to ensure that women can reap the gains of 4IR in the long-term	Only **a third** of graduates from science, technology, engineering, and mathematics (STEM) programs are female
	Lack of inclusive skilling opportunities and social protection mechanisms for flexible workers	Limited focus on social protection policies to address new types of workers that could emerge in a knowledge-based economy, including digital freelancers or gig economy workers

4IR = Fourth Industrial Revolution, AR = augmented reality, ICT = information and communication technology, VR = virtual reality.
Source: Asian Development Bank and AlphaBeta.

B.   Recommendations to Address Challenges

There are several areas in which Uzbekistan can strengthen its approach to 4IR to address the challenges outlined above. This section provides policy recommendations (see Table 8) for policy makers, drawing from international best practices, as summarized in Figure 48.

Figure 48: Relevant Best Practices That Could be Adopted to Tackle Challenges in Adoption of Industry 4.0 Practices

A range of relevant best practices could be adopted to tackle these challenges

Recommendations	Challenges addressed	Examples of countries with best practices/ similar solutions
1 Develop sectoral 4IR adoption plans to coordinate technology adoption and skills development	■ Lack of clearly articulated 4IR vision integrated with jobs and skills aligned across stakeholders ■ Limited alignment between industry and training institutions on current and future skills needs	Singapore
2 Develop programs to guide digital transformation of small and medium businesses	■ Uneven understanding of 4IR technologies amongst firms	Australia, Singapore
3 Strengthen the training capabilities of employers	■ Few training institutions provide relevant 4IR-specific courses on 4IR today	Ireland, Malaysia, Pakistan, Singapore
4 Develop online training platforms	■ Significant changes in skill demand may lead to challenges in hiring workers	Republic of Korea
5 Adopt programs to strengthen industry knowledge of trainers	■ Firms face challenges in hiring graduates that are adequately prepared for their roles	Malaysia, United Kingdom
6 Promote use of innovative technologies to strengthen training delivery	■ Limited use of VR/AR technologies	Australia, Belgium, Singapore, South Africa, United Kingdom
7 Develop targeted programs to ensure that women can benefit from 4IR	■ Job gains from 4IR benefit male workers more than female workers	Australia, Cambodia, Pakistan
8 Develop skilling and labor support programs for digital freelancers	■ Lack of inclusive skilling opportunities and social protection mechanisms for flexible workers	Pakistan

■ Address textile industry challenges   ■ Address construction industry challenges
■ Addresstraining landscape challenges   ■ Address policy assessment challenges

4IR = Fourth Industrial Revolution, AR = augmented reality, VR = virtual reality.
Source: Asian Development Bank (Sustainable Development and Climate Change Department).

Table 8: Summary of Recommendations, Potential Lead Agencies, and Approximate Time Frame for Implementation

Recommendations	Potential Lead Agencies	Approximate Time Frame for Implementation
Develop sectoral 4IR adoption plans to coordinate technology adoption and skills development	Ministry for Innovation Development	12–36 months
Develop programs to guide digital transformation of small and medium businesses	Agency for Development of Small Business and Entrepreneurship, Ministry for Development of Information Technologies and Communications	12–36 months
Strengthen the training capabilities of employers	Ministry of Higher Education in Uzbekistan, Ministry of Employment and Labor Relations	12–36 months
Develop online learning platforms	Ministry of Higher Education	Less than 12 months

continued on next page

Table 8 *continued*

Recommendations	Potential Lead Agencies	Approximate Time Frame for Implementation
Adopt programs to strengthen industry knowledge of trainers	Ministry of Higher Education	12–36 months
Promote use of innovative technologies to strengthen training delivery	Ministry of Higher Education	12–36 months
Develop targeted programs to ensure that women can benefit from 4IR	Ministry for Innovation Development	12–36 months
Develop skilling and labor support programs for digital freelancers	Ministry of Employment and Labor Relations, Ministry for Innovative Development	12–36 months

Source: Asian Development Bank and AlphaBeta.

Recommendation 1: Develop sectoral 4IR adoption plans to coordinate technology adoption and skills development

Uzbekistan has adopted a range of policies to enable firms and workers to transform digitally but could strengthen coordination in implementing these policies. Plans to encourage innovation among firms in Uzbekistan are set out in national strategies, such as the Digital Uzbekistan 2030 road map and Strategy for Innovative Development, 2019–2021. The government periodically announces measures to support firms in specific industries, including textile and garment manufacturers and construction firms, to adopt technology as well as road maps setting out the growth strategy of each industry. In parallel, there are ongoing efforts to make training more relevant to industry needs, with sectoral skills councils established to update professional qualifications standards. However, consultations with government stakeholders suggest that there is a need to align industry development plans and skills development plans more closely. There is currently no single action plan setting out the long-term growth vision for 4IR in each industry that consolidates plans for innovation, technology adoption, and skills development. A clear action plan for 4IR adoption in each industry could strengthen coordination among stakeholders and facilitate long-term planning for firms and training institutions.

The Industry Transformation Maps (ITMs) developed by the Government of Singapore provide a useful model. Industry-specific ITMs have been developed for 23 industries, drawing together the inputs from private and public stakeholders in each industry, including trade associations and key firms. Each ITM charts out the overall growth direction for the industry under different thrusts of transformation, such jobs and skills, productivity, innovation, and internationalization. For example, the Construction ITM sets out Singapore's vision of an advanced and integrated building sector with widespread adoption of leading technologies that can provide quality jobs. Under the productivity and innovation thrusts, it sets out the government's overall goals to support firms to adopt new, advanced technologies and identifies specific technology areas in which innovation will be pursued, for instance cloud and digital technologies, which can connect the entire project delivery process from design, fabrication, to assembly on-site. The jobs thrust sets the objectives of creating quality jobs within the industry and enabling workers to transit smoothly into new, higher-skilled roles enabled by technology adoption. Specific policies and programs are guided by these broad objectives aligned across stakeholders.[15]

---

[15]   Ministry of Trade and Industry. ITMs Construction. https://www.mti.gov.sg/ITMs/Built-Environment/Construction.

The key distinguishing factor between the Singapore ITM model and existing strategies in Uzbekistan is coordination. In Uzbekistan, the digital transformation of firms and workers in the construction industry is guided by a combination of national strategies, government decrees, and the efforts of the sectoral skills councils. In Singapore, industry development, job redesign and skills development efforts are coordinated and guided by the Construction ITM. More importantly, the ITM also guides the long-term direction of government policies related to 4IR in the construction industry, so that firms can plan their technology adoption needs ahead and training institutions know the types of technologies that will be adopted by firms going forward and therefore the corresponding practical knowledge and skills important for graduates.

The Ministry for Innovation Development could take the lead in coordinating similar 4IR action plans in Uzbekistan, working with key line agencies. For instance, an action plan for 4IR technology adoption in the construction industry could be jointly developed with the Ministry of Construction. Policies related to technology adoption, industry development, and skills development often span a wide range of government stakeholders and a strong coordinating agency is needed to coordinate these efforts and resolve any differences between stakeholders; as well as incorporate the needs of the private sector. A clear mandate must therefore be given to the Ministry for Innovative Development. For instance, Singapore's Construction ITM is led by the Building and Construction Authority, the lead agency for the sector, but the overall implementation is overseen by the Future Economy Council chaired by the Deputy Prime Minister.[16]

Recommendation 2: Develop programs to guide digital transformation of small and medium businesses

Understanding and adoption of 4IR technologies vary across firms in Uzbekistan, with small and medium-sized enterprises (SMEs) expected to face more resource and manpower constraints in adopting such technologies. Policies to guide and support the digital transformation of SMEs are critical. The employer surveys revealed that while 35% of construction firms and 53% of textile and garment manufacturers surveyed indicated an advanced understanding of 4IR technologies, around a third of firms in both industries have a limited understanding of 4IR. Research and expert interviews reveal that smaller businesses tend to face more challenges in making the transition to 4IR, including the lack of access to information on tools available and financing, as well as difficulty in attracting workers with the right skill sets (UNIDO 2021). In Uzbekistan, SMEs are the biggest source of employment providing close to 80% of jobs and targeted programs to ensure that they reap the gains of 4IR are critical (OECD 2017).

Australia and Singapore provide some examples of how policy makers can guide and support SMEs in adopting 4IR technologies (see Box 3). In Uzbekistan, the Agency for Development of Small Business and Entrepreneurship and the Ministry for Development of Information Technologies and Communications could take the lead in developing industry digital plans and free or subsidized 4IR-focused advisory programs for SMEs. The development and implementation of these plans and programs could involve local and global technology experts, research institutions, and industry stakeholders such as industry associations. It would be particularly important to work closely with industry associations and business chambers on outreach activities when programs are implemented to ensure that SMEs are aware of the tools available.

---

[16]   Ministry of Trade and Industry. ITMs Overview. https://www.mti.gov.sg/ITMs/Overview.

Box 3: Helping Small and Medium-Sized Enterprises Go Digital in Australia and Singapore

In Singapore, the SMEs Go Digital program supports small and medium-sized enterprises (SMEs) to use digital technologies and build stronger digital capabilities. Various forms of support are available under the program to help firms understanding their digital needs and enable them to meet these needs. One example is the Industry Digital Plans that have been created to guide firms, particularly SMEs, in the adoption of 4IR technologies. The plans provide clear guidance on the industry-specific digital tools that firms can adopt at different stages of growth. Diagnostic tools are available for firms to gauge their readiness to adopt various digital tools (see IMDA Industry Digital Plans). For instance, under the Construction and Facilities Management Industry Digital Plan, construction firms at a nascent stage of technology adoption are guided to implement solutions such as digital wearables for workers to track health and safety information and 3D modeling and immersive visualization technologies. Those at a more advanced stage are guided to implement technologies such as robotics for autonomous construction and facilities management (see IMDA Construction and Facilities Management IDP). The Industry Digital Plans are complemented by other tools under the SMEs Go Digital program. The Productivity Solutions Grant (PSG) supports firms keen on adopting IT solutions and equipment to enhance business processes by subsidizing the cost of adopting a suite of SME-friendly approved solutions (see Enterprise Singapore Productivity Solutions Grant). The SME Digital Tech Hub provides expert advice to SMEs on how they can transform their businesses using digital technologies (see IMDA SMEs Go Digital).

In Australia, the Digital Solutions – Australian Small Business Advisory Services program provides independent advice to Australian small businesses to help them build their digital capabilities. The first interaction with the service is free and small businesses with fewer than 20 full-time employees, as well as sole traders, can access consultancy services at a subsidized rate of A$44 for 7 hours of support. Specific advice is offered in areas such as how digital tools can help the business; e-commerce; social media and digital marketing; as well as cybersecurity and data privacy (see Business.gov.au. Digital Solutions - Australian Small Business Advisory Services).

Sources: IMDA. *Industry Digital Plans.*https://www.imda.gov.sg/programme-listing/smes-go-digital/industry-digital-plans; IMDA. *Construction and Facilities Management IDP.* https://www.imda.gov.sg/programme-listing/smes-go-digital/industry-digital-plans/Construction-and-Facilities-Management-IDP; Enterprise Singapore. *Productivity Solutions Grant.* https://www.enterprisesg.gov.sg/financial-assistance/grants/for-local-companies/productivity-solutions-grant; IMDA. *SMEs Go Digital.* https://www.imda.gov.sg/programme-listing/smes-go-digital; Business.gov.au. Digital Solutions - Australian Small Business Advisory Services https://business.gov.au/expertise-and-advice/digital-solutions-australian-small-business-advisory-services.

Recommendation 3: Strengthen the training capabilities of employers

A large proportion of employers in Uzbekistan provide OJT to their employees. Programs to support employers to strengthen their training capabilities—both in-house and through training providers—could help to strengthen the effectiveness of training provided. In addition, such programs could also address the specific resource constraints faced by SMEs in providing training.

The employer surveys revealed that while a significant proportion of employers in both the textile and garment manufacturing industry do not find graduates to be adequately prepared for their jobs, most feel that they provided sufficient training for their graduates. OJT is the preferred training channel. Employers in both industries currently provide around 70% of workers with OJT. With the adoption of 4IR rapidly changing the skill needs of employers, and possibly the types of equipment used, training institutions could find it challenging to provide equipment-specific training to large cohorts of students or in training students in niche skills required by employers. The strengthening of employers' training capabilities could help to bridge this gap.[17] While most firms feel that they have sufficient training capabilities in 2020, this could be due to the relatively low level of skills required in some firms, particularly those with a limited understanding of 4IR technologies. In addition, the quality of training could differ between firms, with SMEs that employ around 80% of Uzbekistan's workforce typically having fewer resources to provide high-quality training based on expert interviews (OECD 2017).

To ensure that employer-provided training is relevant and able to help workers meet the changing demands of 4IR, the Ministry of Higher Education in Uzbekistan could work closely with the Ministry of Employment and Labor Relations to devise programs to strengthen the in-house training capabilities of larger employers and recognize credentials obtained through such training. These could complement plans to establish a dual education system that allows workplace-based training to supplement classroom teaching (*UZ Daily* 2021c). Specific to SMEs, the Agency for Development of Small Business and Entrepreneurship could develop programs to support SMEs to augment their training capabilities through external training providers (see Box 4). The Ministry of Employment and Labor Relations has established adult learning employment centers around Uzbekistan, and these also could be leveraged to provide training.

Box 4: Programs to Strengthen Employers' Training Capabilities in Ireland, Malaysia, Pakistan, and Singapore

Existing programs in Pakistan and Singapore support firms to strengthen their training capabilities. In Pakistan, the Punjab Skills Development Fund's (PSDF) Industry Training Programs provide support to businesses to train youths, to ensure that the training curriculum is agile and in line with the emerging skills needs of businesses. Under this initiative, businesses design the training coursework and teaching materials, select trainers to teach the course and conduct the course at their own facilities or in partnership with other training providers. The PSDF funds the training and pays trainees monthly stipends to incentivize them to complete the course but requires a dedicated classroom and instructor to ensure that proper training is carried out. It also requires employers to commit to hiring 50% of the subsidized trainees (see Punjab Skills Development Fund webpage). In Singapore, the Institute of Technical Education (ITE) provides consultancy services to set up structured on-the-job-training (OJT) programs according to the pedagogic competencies and needs of organizations. Organizations that provide training to employees under approved OJT programs are also eligible for government grants (see ITE Industry Training Schemes).

Programs in Ireland and Malaysia recognize the resource constraints faced by small and medium-sized enterprises (SMEs) in providing training for their workers. In Ireland, the Skillnet Business Networks support businesses, to identify the skills needed to sustain their growth. The design, sourcing, and delivery of training courses is coordinated by the Skillnet Business Network, so that SMEs which have limited resources do not have to take on the administrative burden of planning the courses. The cost of training is also subsidized by Skillnet Ireland (see Skillnet Ireland Webpage). In Malaysia, a Skills Upgrading Program provides grants covering 70% of training fees for technical and soft skills for SMEs to train their workers (see SME Corp Malaysia webpage).

Sources: Punjab Skills Development Fund. *Industrial Training Programme*. https://www.psdf.org.pk/industrial-training-programmes/; ITE. *Industry Training Schemes*. https://www.ite.edu.sg/employers/industry-training-schemes/certified-on-the-job-training-centre; Skillnet Ireland. *Our Support for Businesses*. https://www.skillnetireland.ie/about/our-support-for-smes/; SME Corp Malaysia. *Skills Upgrading Programme*. http://www.smecorp.gov.my/index.php/en/slides/86-program-sme/103-skills-upgrading-programme.

---

[17]   Consultations with sectoral experts

Recommendation 4: Develop online learning platforms

The adoption of 4IR technologies will constantly change the skills required of workers and it is critical to put in place frameworks and platforms for workers to continually reskill. In Uzbekistan, a National Qualifications Framework has been established and there are plans to establish competency assessment centers to recognize the results of nonformal learning and award professional qualifications to workers. Outside these plans however, opportunities for the recognition of qualifications acquired outside formal education (e.g., short-term training courses offered by online providers) remains limited (ETF 2021). Adult learning opportunities that lead to recognized qualifications are largely limited to courses offered by vocational training institutions and there is a limited focus on 4IR technologies or soft skills (ETF 2021). Uzbekistan could consider online-learning platforms to rapidly build up the new skills required by employers.

Under the Republic of Korea's Life-Long Learning Promotion Plan (2018–2022), online learning platforms have been established to upskill the population in a range of areas, including digital skills. A key platform is the Korean massive Open Online Courses (K-MOOC), which, since its launch in 2015, has developed over 1,700 accredited courses at the higher education level through partnerships with local universities, with a significant share focused on advanced digital courses such as machine learning, AI navigation and perception, and mathematics for data scientists.[18] In addition, the Ministry of Education runs a Distance University Education program, in which workers have the option to take university courses in new ICT skills, and upon completion of such programs, are awarded degrees.[19]

Uzbekistan could work with local universities and technical training institutions to launch platforms for open online courses or distance learning degrees. The Ministry of Higher Education coordinates the national TVET system and engages industry stakeholders to ensure that training curricula is relevant to industry needs. The Ministry could play a key role in developing and accrediting online courses, particularly courses focused on 4IR-related skills. The One Million Uzbek Coders program led by the Dubai Future Foundation provides an example of how online training programs could be effective in Uzbekistan. It provides free distance learning courses in data analytics and computer programming, with online certificates from the program paving the way to a nanodegree recognized by global IT employers (*Gulf News* 2021). The program attracted over 500,000 trainees keen to pursue ICT careers. Similar nationwide initiatives focused on a larger range of skill areas could be effective in helping Uzbekistan's workforce develop future skills.

Recommendation 5: Adopt programs to strengthen industry knowledge of trainers

The quality of instructors is integral to any training and education system, and programs to strengthen the industry knowledge of trainers can help to ensure that graduates gain relevant and practical skills needed to succeed in the workplace. In Uzbekistan, the employer surveys revealed that employers in the construction and textile and garment manufacturing industries face challenges in identifying and hiring graduates who are sufficiently prepared for the job by their education or training. Consultations with government stakeholders further revealed the need for training curricula to be more closely aligned with industry's technology adoption plans, and suggest a need for closer collaboration between industry and training institutions on the delivery of training, to ensure that graduates have the practical skills required to take on jobs.

18   K-MOOC. Courses. http://www.kmooc.kr/courses.
19   Government of the Republic of Korea, Ministry of Education. *Lifelong Education*. http://english.moe.go.kr/sub/info.do?m=020107&s=english.

As such, Uzbekistan could consider programs to improve the industry knowledge of teaching staff in training institutions, referencing best practices in Malaysia and the United Kingdom (see Box 5). The Ministry of Higher and Secondary Specialized Education could take the lead in programs to ensure that instructors not only have strong pedagogical knowledge, but also a good grasp of current industry trends and practical skills sought after by employers.

Box 5: Programs to Strengthen Industry Knowledge of Trainers in Malaysia and the United Kingdom

To ensure that instructors have updated skills and knowledge, industry stakeholders and training providers could collaborate on industrial attachments for instructors or encourage experienced industry professionals to join their technical and vocational education and training workforce.

In Malaysia, INTI International University & Colleges created the Faculty Industry Attachment program. The program enables teaching staff to work with industries as part of their regular working hours to broaden their practical experiences and stay abreast of the latest developments in the industry. Lecturers undergo an industrial attachment in related organizations for up to 96 hours, an estimate of 2.5 working weeks, to gain an in-depth understanding of the current demands of the industry (INTI International University & Colleges 2019).

In the United Kingdom, the Taking Teaching Further program is a national initiative to attract experienced industry professionals with expert technical knowledge and skills to work in education. Training providers receive funding support to recruit people with industry experience to retrain for teaching positions. The funding covers the cost of undertaking a teaching qualification as well as the costs of inducting a new teacher into the system (i.e., work-shadowing arrangements and reduced workload for new teachers) (see Education and Training Foundation Taking Teaching Further homepage).

Sources: INTI International University & Colleges. 2019. Industry *Attachment Programme Redefines Teaching at INTI*. https://newinti.edu.my/industry-attachment-programme-redefines-teaching-at-inti/; Education and Training Foundation. *Taking Teaching Further*. https://www.et-foundation.co.uk/supporting/support-for-teacher-recruitment/taking-teaching-further/.

Recommendation 6: Promote the use of innovative technologies to strengthen training delivery

Innovative technologies such as AR, VR, or AI can help to strengthen training delivery in Uzbekistan and enable training institutions to better prepare graduates for the workforce. The surveys conducted found that 56% of training institutions in Uzbekistan use online self-learning modules to deliver training and only a third use VR or AR mechanisms. There is scope for training quality to be strengthened through improved training approaches enabled by 4IR technologies (see Box 6).

In Uzbekistan, collaboration with the private sector coordinated by the Ministry of Higher Education could help to address potential resource constraints faced by training institutions in the adoption of 4IR technologies. Possible policy levers include pilot courses jointly developed by industry and training institutions to test the effectiveness of new training technologies.

Box 6: Use of Artificial Intelligence, Virtual Reality, and Augmented Reality Technologies to Improve Learners' Experience in India, South Africa, and the United States

Artificial intelligence (AI) technology can be used to develop customized training materials for each student based on their ability, preferred mode of learning, and pace of learning (Schmelzer 2019). In South Africa, the Department of Basic Education rolled out Ms Zora, an AI-based educational platform, to support the introduction of coding and robotics curriculum in schools. The AI-powered virtual assistant serves as both a teacher's assistant and personal tutor to pupils across all grades, enabling students to self-learn (iWeb 2020).

In the United States, schools are adopting augmented reality (AR) and other immersive technologies to build up soft skills such as creative problem solving and observation skills through different types of AR environments (sources: zSpace and YourStory). In India, EdTech start-up fotonVR develops virtual reality (VR) content for schools, providing a turn-key solution that includes the hardware, setting up the classroom, software, content, and training to the teachers. The cost of setting up a classroom, along with the content, kits, and training to teachers ranges from approximately $8,000 to $20,000 (₹600,000 to 1.6 million) per classroom. Once the setup is done, fotonVR charges approximately $1,350–$2,700 (between ₹100,000 to 200,000) as an annual subscription charge (Your Story 2020).

Virtual reality (VR) technologies also have specific applications to train students in some industries. In the information technology-business process outsourcing industry for instance, VR technologies can enable agents to experience realistic customer service scenarios that test and develop their capabilities, without exposing actual customers to agents who are not fully trained (Saeb 2017). In the industrial manufacturing space, German firm Siemens uses AR to help trainees practice stimulated welding (Garage 2019, HR Technologist 2019).

Industry 4.0 technologies can also be adopted by institutions to support students in their professional and personal development. For instance, a job interview simulator created by soft skills training specialist Bodyswaps teaches users various interview techniques and allows them to practice answering questions confidently, while using behavioral analytics to assess the user's verbal and nonverbal performance (Graham 2021). Such technologies could address problems such as a lack of trained manpower at training institutions for dedicated career or interview coaching.

Sources: R. Schmelzer. 2019. AI Applications in Education. *Forbes*. 12 July; iWeb. 2020. Back to School: Robotics, Coding Curriculum Pushed Back. 15 January; zSpace. *Beyond STEM: Building Soft Skills with Augmented and Virtual Reality*; Your story. 2020. *This Edtech Startup is Using Virtual Reality-Based Content to Have Real-Life Impact on Students*; P. Saeb. 2017. *Virtual Reality Potential for Training Contact Centre Agents*. contact-centres.com. 3 January; R. Garage. 2019. *Use Cases of Augmented Reality in Education and Training*. 18 January; HR Technologist. 2019. *How AR and VR are Revolutionizing Soft Skills Training in 2019*. 22 April; P. Graham. 2021. *Don't Stress About a Job Interview with Bodyswaps' New Simulator*. VR Focus. 27 May.

Recommendation 7: Develop targeted programs to ensure that women can benefit from 4IR

The employer surveys show that the adoption of 4IR technology will lead to a higher proportion of technical jobs in the workforce, and digital and/or ICT skills will become more important. Targeted programs are needed to overcome sociocultural norms in Uzbekistan and ensure that female workers can benefit from higher-skilled, better-paying technical jobs created by 4IR in the long-term. An ILO study found that while reforms are underway to promote education and training for women and girls, a more comprehensive approach to gender equality in education is needed. For instance, current training programs aimed at women tend to focus on traditional skills such as sewing, cooking, and hairdressing rather than ICT or technical fields. Currently, while 45% of male tertiary students in Uzbekistan pursue STEM fields, a similar percentage of female workers are enrolled in education-related disciplines. Policy makers in Uzbekistan can consider initiatives to build stronger digital literacy among women; as well as encourage female participation in STEM (ILO 2021).

Cambodia and Pakistan provide examples of programs aimed at building stronger digital literacy among women. In Pakistan, the ICTs for Girls program provides opportunities for girls and women to improve their digital literacy and employability (*ITU News* 2018). As part of this program, tens of thousands of girls and women from disadvantaged segments of society are provided digital infrastructure with state of art machines in fully broadband supported environments. In Cambodia, the United Nations Educational, Scientific and Cultural Organization partnered the Garment Manufacturers Association in Cambodia to support lifelong learning among garment factory workers across Cambodia under the Factory Literacy Program (*The Phnom Penh Post* 2021), wherein female factory workers are equipped with functional literacy and numeracy skills. Those who pass the course's final exam obtain a certificate of completion accredited by the Ministry of Education, Youth and Sport. Cambodia's experience could be particularly relevant for Uzbekistan's textile and garment manufacturing industry where female manual workers form a significant proportion of the workforce. While the study estimates that female manual workers could initially benefit from net job gains created by 4IR adoption as the industry grows rapidly, it also cautions that some manual jobs could eventually be displaced as the industry moves toward a higher proportion of technical roles to remain competitive in the long-term.

In Australia, there are various programs focused on encouraging STEM pursuits for female students. The Curious Minds program funded by the Department of Education, Skills and Employment is aimed at highly capable female students (aged around 15–16 years old) who have an interest in STEM. This is a 6-month program that combines two residential camps and a coaching program to help ignite girls' passion in STEM. The Government of Australia also funded the development of the Girls in STEM Toolkit, which contains resources for female students to find out more about STEM careers; and for teachers to inspire and encourage girls to feel confident and enthusiastic about STEM-related jobs (Government of Australia 2019). Policy makers in Uzbekistan could consider similar initiatives to strengthen the capabilities and interest of women in taking on technical professions so that they will benefit from the job gains from 4IR. This could be done in collaboration with international partners, such as the ILO or the Asian Development Bank. The Ministry for Innovation Development could coordinate efforts to ensure that female workers can also benefit from 4IR.

Recommendation 8: Develop skilling and labor support programs for digital freelancers

One key aspect of the digital economy and 4IR is the proliferation of digital freelancers. There is significant potential for Uzbekistan workers to find quality jobs as digital freelancers with the rise of the global freelance economy against the backdrop of the COVID-19 pandemic. A study by Upwork revealed that amid the pandemic, over a third of the American workforce did freelance work, contributing $1.2 trillion to the US economy.

Policy makers in Uzbekistan recognize the potential of the digital freelancing economy. The Future Skills Uzbekistan program, which aims to increase the number of qualified IT specialists through a comprehensive training program covering areas such as mobile application development, IT security and network technologies, and 3D modeling includes a component to support trainees to find digital freelancing jobs (*UZ Daily* 2021d). To provide further support for digital freelancers, Uzbekistan could reference Pakistan's efforts to create a conducive environment for freelance digital work through the National Freelancing Facilitation Policy 2021 (see Box 7).

Tapping on Pakistan's example, the Ministry of Employment and Labor Relations and Ministry for Innovative Development, in particular, could lead efforts to create an enabling environment for digital freelancers and remove potential barriers surrounding taxes and fees related to transfer of foreign currencies across borders, as most clients are likely to be based outside of Uzbekistan.

Box 7: Various Forms of Government Support for Digital Freelancers in Pakistan

In the region, Pakistan is home to the world's fourth largest number of digital freelancers working on online platforms for contractual jobs (*Geo News* 2021). The Pakistan Software Export Board (PSEB) estimates that the exponential growth in the number of freelancers in Pakistan in recent years has resulted in a revenue of $150 million earned by the freelancers in 2019. Government agencies are keen to support such growth and the Punjab Information Technology Board has launched e-Rozgaar centers to provide training for youths in areas such as e-commerce and website development that will enable them to work as digital freelancers (see e-Rozgaar homepage).

The Pakistan Ministry of Information Technology and Telecommunication has drafted the National Freelancing Facilitation Policy 2021 to build the number of active freelancers in Pakistan to 1 million (Government of Pakistan, Ministry of Information Technology and Telecommunication 2021). The policy includes programs to develop the skills of local freelancers such as the launch of new training and technology certifications initiatives for freelancers by the Pakistan Software Export Board and fiscal incentives such as income tax holidays and subsidized health and life insurance for digital freelancers.

Sources: *Geo News*. 2021. 47% Growth Seen in Pakistan's IT Exports from July–May in Fiscal Year 2020-21. 26 June. https://www.geo.tv/latest/357011-47-growth-seen-in-pakistans-it-exports-from-july-may-in-fiscal-year-2020-21; M. Qasim. 2021. *$150 Million Revenue brought by "Pakistani Freelancers" in a Year*. https://startuppakistan.com.pk/150-million-revenue-brought-by-pakistani-freelancers-in-a-year/; e-Rozgaar. About us. https://www.erozgaar.pitb.gov.pk/#erb04; Government of Pakistan, Ministry of Information Technology and Telecommunication. 2021. National Freelancing Facilitation Policy 2021 Consultation Draft. Islamabad. https://moitt.gov.pk/SiteImage/Misc/files/National%20Freelancing%20Facilitation%20Policy%202021%20-%20Consultation%20Draft%202_0.pdf.

C. Industry-Specific Priorities

While these recommendations apply to both the textile and garment manufacturing and construction industries, a set of priorities unique to each industry should be considered when implementing the respective recommendations.

Textile and Garment Manufacturing Industry

Among the textile and garment manufacturing firms surveyed, 71% indicated a good understanding of 4IR technologies and more than half agreed that companies in their supply chain have adopted such technologies. However, it appears that training institutions face challenges in offering sufficient 4IR-related courses. For instance, while 55% of textile and garment manufacturing firms have adopted autonomous robots, only 10% of training institutions offer relevant courses. This challenge is likely to be exacerbated as 4IR changes the task profiles of jobs in the industry and skills needs. Creative thinking and/or design and digital and/or ICT skills will become the most valued skills by employers by 2025, with 4IR adoption and a substantial amount of reskilling and upskilling for workers would be needed, particularly for digital and/or ICT skills. The development of sectoral 4IR adoption plans to coordinate technology adoption and skills development is therefore particularly critical for the textile and garment manufacturing industry to ensure that the curricula and courses provided by training institutions are aligned with the rapidly changing skills needs of employers. At the same time, programs to strengthen the industry knowledge of trainers would also be necessary to ensure that trainers are equipped to deliver courses in these new skills areas and technologies. These efforts could be complemented by programs to strengthen the training capabilities of employers given that textile and garment manufacturing firms currently provide OJT to 70% of their employees.

Table 9: Summary of Findings in the Textile and Garment Manufacturing Industry

Key Findings

Potential job displacement (% of current work force): 147,000 (43%)
Potential job gains (% of current work force): 247,000 (71%)
Net job gains from 4IR (% of current work force): 100,000 (29%)
Top three in-demand skills in 2025: Creative thinking and/or design, digital and/or ICT skills, social and interpersonal skills

Key Challenges	Findings	Recommendations
The adoption of 4IR will rapidly change task profiles and skills needs in the textile and garment manufacturing industry	Creative thinking and/or design and digital and/or ICT skills will become the most valued skills by employers by 2025, with 4IR adoption	Develop sectoral 4IR adoption plans to coordinate technology adoption and skills development
Training institutions do not offer courses in 4IR technologies relevant to textile and garment manufacturing employers	Although **55%** of textile and garment manufacturing firms have adopted autonomous robots, only **10%** of training institutions offer relevant courses	Strengthen the training capabilities of employers
Firms face challenges in hiring graduates that are adequately prepared for their roles	Only **47%** of textile and garment manufacturers agree that graduates hired in the past year were adequately prepared for the job	Adopt programs to strengthen industry knowledge of trainers

4IR = Fourth Industrial Revolution.
Source: Asian Development Bank (Sustainable Development and Climate Change Department).

Construction Industry

Firms in the construction industry vary widely in their understanding of 4IR with around half of firms surveyed indicating a limited understanding of 4IR technologies while 35% reported an advanced understanding (Table 10). Interviews with in-market experts suggest that usage of 4IR technologies in Uzbekistan's construction industry is limited to larger firms involved in large-scale infrastructure projects, while smaller firms (e.g., subcontractors) see less need for such technologies and lack the resources to deploy them. In addition, most advanced technologies and systems are imported as domestic innovation is at a fledging stage, which means that firms need to expend resources to adapt these systems to local conditions and train their workers to use them, putting smaller firms with less resources at a further disadvantage. Programs to build a strong awareness of digital tools available will therefore be particularly critical for small firms in the construction industry. In addition, programs to expand the use of AR/VR technologies and increase the share of women in technical roles could be particularly useful for Uzbekistan's construction industry. VR-based training is particularly relevant for industries such as construction, in which untrained workers face a high risk of accidents but only 33% of training institutions use VR/AR technologies to deliver training as of 2020.

Table 10: Summary of Findings in the Construction Industry

Key Findings

Potential job displacement (% of current work force): 556,000 (42%)
Potential job gains (% of current work force): 890,000 (67%)
Net job gains from 4IR (% of current work force): 334,000 (25%)
Top three in-demand skills in 2025: Creative thinking and/or design, digital and/or ICT skills, critical thinking

Key Challenges	Findings	Recommendations
Understanding of 4IR technologies varies significantly between firms	**35%** of construction firms have an advanced understanding of 4IR technologies, but a similar proportion have not heard of 4IR	Develop programs to guide digital transformation of small and medium businesses
Limited use of VR/AR technologies in the classroom despite the potential for such technologies for construction-related training	Only **33%** of training institutions use VR/AR technologies to deliver training	Promote use of innovative technologies to strengthen training delivery
Job gains from 4IR benefit male workers significantly more than female workers	The number of jobs expected to be gained by male workers exceeds the number of jobs expected to be gained by female workers by **6.5x**	Develop targeted programs to ensure that women can benefit from 4IR

4IR = Fourth Industrial Revolution.
Source: Asian Development Bank (Sustainable Development and Climate Change Department).

APPENDIX
Participants in the National Consultations

Table A1: Stakeholders Engaged in Initial Consultations for Uzbekistan

No.	Name	Designation	Organization
Government Ministries and Agencies			
1	Oybek Shagazatov	Head, Main Department of Cooperation with International Financial Organisations	Ministry of Investments and Foreign Trade
2	Abdullo Makhmudov	Deputy Head	Ministry of Finance
3	Mirzoxidov Miraziz Mirzoxid Oglu	Head, Department for Monitoring the Formation and Implementation of Localization Projects, Uzbektelecom	Ministry for Development of Information Technologies and Communications
4	Izbasarov Umar Agzamovich	Deputy Head, Strategic Development and Investment Projects Department, Uzbektelecom	
5	Hikmatullaev Timur Shukhrattulaevich	Chief Specialist, Department for Development of International Relations in Science and Innovation	Ministry of Innovational Development
6	Ismailov Islam	Deputy Head, International Cooperation and Interaction with the ILO	Ministry of Employment and Labor Relations
7	Saparov Alexey	Project Manager, Center for Labor Project Management	
8	Asliddin Annaev	Officer	
9	Sitov Kadambay Geldibaevich	Head, Department for Integration of Production and Education	Ministry of Public Education
10	Magay Leonid Svyatoslavovich	Deputy Head, Human Resources Department	
11	Abdihalil Rakhmonov	Head	National Institute for Labor Market Research
12	Allabergenov Anvar Alimbaevich	Head, Department for the Development of Educational Standards of TVET	Ministry of Higher and Secondary Specialized Education
13	Sharofaddinov Shikhnazar Anvarovich	Deputy Director, Institute of Pedagogical Innovations	
Nongovernment Stakeholders			
14	Isaev Makhmudjon Nimatovich	Representative	Federation of Trade Unions of Uzbekistan
15	Yusupova Shakhnoza	Representative	Association of Businesswomen of Uzbekistan

continued on next page

Table A1 *continued*

No.	Name	Designation	Organization
16	Natalya Timofeeva	Head, Personnel Department	Monocenter to provide services to the unemployed population "Ishga Marhamat" LLC
17	Dusheboev Fakhriddin	Deputy Director for Academic Affairs	Center for vocational training of unemployed citizens in the city of Tashkent (for training of displaced workers)
18	Juraev Rustam	Analysis and Monitoring Specialist, International Department	IT Park Uzbekistan
19	Daskaev Bekir	Resident Relations Specialist, International Department	
20	Dildora Abidjanova	National Program Officer	Embassy of Switzerland in Uzbekistan

ILO = International Labour Organization, TVET = technical and vocational education and training.
Source: Asian Development Bank and AlphaBeta.

Table A2: Stakeholders Engaged in Further Consultations for Uzbekistan

No	Government Agency	Representative Name	Designation
1	Ministry of Investments and Foreign Trade	Oybek Shagazatov	Head, Main Department of Cooperation with International Financial Organizations
2	Ministry of Finance	Makhmudov Abdullo	Deputy Head, Department of Cooperation with International Financial Organizations
3	Ministry of Employment and Labor Relations	Odil Kamiljanov	Head, Vocational education sector
		Ismoilov Islam Ilhomovich	Deputy Head, International Relations and ILO Relations Department
		Mahmudov Temur Asomiddinovich	Head, Department for Employment of Unemployed Population
		Shomaqsud Shoabdukarimov	Head, Department of Vocational Training and Development of Professional Skills
4	Institute of Pedagogical Innovation for TVET	Shikhnazar Sharofiddinov	Deputy Director
5	Ministry of Higher and Secondary Specialized Education	Alijanov Utkir	Head, Main Directorate of Educational and Methodological Coordination of Professional Education
		Anvar Allabergenov	Head, Department for the Development of Educational Standards and Qualification Requirements
		Nodir Ergashev	Chief Specialist, Investment Projects Implementation Department
		Rakhmonov Utkir	Chief Specialist, Department for the Development of Research and Innovation Activities

continued on next page

Table A1 *continued*

No	Government Agency	Representative Name	Designation
6	Ministry of Public Education	Rustamov Elmurad Dilmuradovich	Head, Investment Promotion and Monitoring Department
		Shaymardanov Alkham Muradovich	Chief Specialist, Department of Investment Promotion and Monitoring
7	Ministry of Construction	Saydamov Dilmurod Bahriddinovich	Department for Working with Education Establishments
8	Ministry of Economic Development and Poverty Reduction of Uzbekistan	Rakhimov Farkhad Khushbakovich	Head, Department for Development of Innovative Technologies and Science

ILO = International Labour Organization.
Source: Asian Development Bank and AlphaBeta.

References

ACT/EMP and ILO. 2017. ASEAN in Transformation: How Technology is Changing Jobs and Enterprises. *Cambodia Country Brief.* https://www.ilo.org/actemp/publications/WCMS_579672/lang--en/index.htm.

AlphaBeta. 2017. *The Automation Advantage.* https://alphabeta.com/wp-content/uploads/2017/08/The-Automation-Advantage.pdf

Armstrong, M. M. 2020. Cheat Sheet: What is Digital Twin? *IBM Business Operations Blog.* 4 December. https://www.ibm.com/blogs/internet-of-things/iot-cheat-sheet-digital-twin/.

Asian Development Bank (ADB). 2020a. *Pakistan: Reviving Growth through Competitiveness.* Manila. https://www.adb.org/publications/pakistan-reviving-growth-through-competitiveness.

———. 2020b. Uzbekistan: *Quality Job Creation as a Cornerstone for Sustainable Economic Growth.* Manila. https://www.adb.org/publications/uzbekistan-job-creation-economic-growth.

Association for Advancing Automation. 2018. Construction Robots Will Change the Industry Forever. 17 April. https://www.automate.org/blogs/construction-robots-will-change-the-industry-forever.

Barber. 2007. *Instruction to Deliver: Fighting to Transform Britain's Public Services.* https://books.google.com.sg/books/about/Instruction_to_Deliver.html?id=MLcbAQAAMAAJ&redir_esc=y

Bournemouth University. Using Virtual Reality to Train Nurses. https://www.bournemouth.ac.uk/why-bu/fusion/using-virtual-reality-train-nurses.

Buehler, M., Buffet, P. P., and Castagnino, S. 2018. The Fourth Industrial Revolution is About to Hit the Construction Industry: Here's How It Can Thrive. *World Economic Forum.* 13 June. https://www.weforum.org/agenda/2018/06/construction-industry-future-scenarios-labour-technology/.

Built Worlds. 2017. Cloud Adoption is Rising in the AEC—But What Are We Missing? Sage Weighs In. 29 June. https://builtworlds.com/news/cloud-adoption-is-rising-in-the-aec-but-what-are-we-missing-sage-weighs-in/.

Business.gov.au. Digital Solutions – Australian Small Business Advisory Services. https://business.gov.au/expertise-and-advice/digital-solutions-australian-small-business-advisory-services.

Buyuk Kelajak. 2019. *Uzbekistan 2035.* Tashkent. https://uzbekistan2035.uz/uzbekistan-2035/.

CB Insights. 2021. The Future of Fashion: from Design to Merchandising, How Tech is Reshaping the Industry. 11 May. https://www.cbinsights.com/research/fashion-tech-future-trends/.

Chea, T. 2021. 3D Printing's New Challenge: Solving the US Housing Shortage. *AP News*. 28 April. https://apnews.com/article/health-technology-lifestyle-business-homelessness-5ae52b49708f2e29a4c5c2cbb7e1e09d.

Clarion Technology. How IOT Transforms the Way to a More Sustainable Textile Manufacturing. https://www.clariontech.com/blog/how-iot-transforms-the-way-to-a-more-sustainable-textile-manufacturing.

Del Buono, A. A. 2018. Combining Smart Technologies for High Quality. *Control Global*. 29 October. https://www.controlglobal.com/articles/2018/combining-smart-technologies-for-high-quality/.

Education and Training Foundation. Taking Teaching Further. https://www.et-foundation.co.uk/supporting/support-for-teacher-recruitment/taking-teaching-further/.

e-Rozgaar. About Us. https://www.erozgaar.pitb.gov.pk/#erb04.

Enterprise Singapore. Productivity Solutions Grant. https://www.enterprisesg.gov.sg/financial-assistance/grants/for-local-companies/productivity-solutions-grant.

European Commission. 2020. *A Report of the ET 2020 Working Group on Vocational Education and Training (VET)*. Brussels. https://ec.europa.eu/social/BlobServlet?docId=23274&langId=en.

European Training Foundation (ETF). 2021. *Torino Process 2018–2020 Uzbekistan National Report*. Turin. https://openspace.etf.europa.eu/sites/default/files/2021-03/TRPreport_2020_Uzbekistan_EN.pdf.

Fibre2Fashion. 2020. Uzbek Textile-Garment Exports Rise by 112% from Jan to Jul. 1 September. https://www.fibre2fashion.com/news/textile-news/uzbek-textile-garment-exports-rise-by-112-from-jan-to-jul-269570-newsdetails.htm.

———. 2021 30-35% of Vietnam's Textile-Garment Operations on Hold Due to COVID-19. 5 August. https://www.fibre2fashion.com/news/apparel-news/30-35-of-vietnam-s-textile-garment-operations-on-hold-due-to-covid-19-275592-newsdetails.htm.

FutureCIO. 2020. Alternative Reality Finds Real-World Use Cases in Construction. 19 November. https://futurecio.tech/alternative-reality-finds-real-world-use-cases-in-construction/.

Garage, R. 2019. Use Cases of Augmented Reality in Education and Training. 18 January. https://rubygarage.org/blog/augmented-reality-in-education-and-training.

Geo News. 2021. 47% Growth Seen in Pakistan's IT Exports from July-May in Fiscal Year 2020-21. 26 June. https://www.geo.tv/latest/357011-47-growth-seen-in-pakistans-it-exports-from-july-may-in-fiscal-year-2020-21.

George Lawton. 2021. Digital Twins Help Transform the Construction Industry. Venture Beat. 18 June. https://venturebeat.com/2021/06/18/digital-twins-help-transform-the-construction-industry/.

Gillet, K. 2021. XYZ Reality Raises £20m to Build AR for Construction Industry. Sifted. 14 June. https://sifted.eu/articles/xyz-reality-20m-constructiontech/.

Global Construction Review. 2016. Asian Firms Turn to Gaming Technology to Ramp up Safety Training. 7 April. https://www.globalconstructionreview.com/asian-firms-turn-gaming-tech7nology-ra7mp-sa7fety/.

Google Cloud Blog. 2021. How to Lower Costs and Improve Innovation with Cloud Computing. 27 July. https://cloud.google.com/blog/topics/research/how-to-lower-costs-and-improve-innovation-with-cloud-computing.

Government of Australia. 2019. *Australian Government Science, Technology, Engineering and Mathematics (STEM) Initiatives for Girls and Women*. Canberra. https://www.industry.gov.au/sites/default/files/March%202020/document/australian-government-stem-initiatives-for-women-and-girls-2019.pdf.

Government of the Republic of Korea, Ministry of Education. Lifelong Education (in Korean). http://english.moe.go.kr/sub/info.do?m=020107&s=english.

Government of Pakistan, Ministry of Information Technology and Telecommunication. 2021. *National Freelancing Facilitation Policy 2021 Consultation Draft*. Islamabad. https://moitt.gov.pk/SiteImage/Misc/files/National%20Freelancing%20Facilitation%20Policy%202021%20-%20Consultation%20Draft%202_0.pdf.

Government of Uzbekistan. 2018. *Presidential Decree About Approval of Strategy of Innovative Development of the Republic of Uzbekistan for 2019–2021* (translated from the Uzbek language). Tashkent. https://cis-legislation.com/document.fwx?rgn=109926.

———. 2019a. *Education Sector Plan of Uzbekistan 2019–2023* (translated from the Uzbek language). Tashkent. https://www.globalpartnership.org/content/education-sector-plan-2019-2023-uzbekistan.

———. 2019b. *Presidential Decree on Approval of the Concept of Development of the Higher Education System of Uzbekistan Until 2030* (translated from the Uzbek language). Tashkent. http://www.erasmusplus.uz/images/shared/file/Concept%20of%20Higher%20Education%20of%20Uzbekistan%20until%202030_EN%20_NEO%20UZ.pdf.

———. 2020a. *Presidential Decree on Approval of the Strategy "Digital Uzbekistan 2030" and Measures for its Effective Implementation* (translated from the Uzbek language). Tashkent. https://lex.uz/docs/5030957#-5031880 https://uzdon.uz/uz/news/info/uzbekistan/165/.

———. 2020b. *Presidential Decree About Approval of Strategy of Modernization and Innovative Development of the Construction Industry for 2021–2025* (translated from the Uzbek language). Tashkent. https://cis-legislation.com/document.fwx?rgn=128834#A5XB0PVLCR.

———. 2020c. *Resolution of the President on Measures for the Cardinal Enhancement of the System of Assessment of Qualifications and Providing the Labor Market with Qualified Personnel* (translated from the Uzbek language). Tashkent. https://cis-legislation.com/document.fwx?rgn=129993.

———. 2020d. *Law about Employment of the Population* (translated from the Uzbek language). Tashkent. https://cis-legislation.com/document.fwx?rgn=128048.

———. 2020e. *Presidential Decree about Measures for Urgent Support of the Textile and Sewing and Knitted Industry* (translated from the Uzbek language). https://cis-legislation.com/document.fwx?rgn=124502.

———.2020f. *Presidential Decree About Measures for Development of Education and Education, and Science During the New Period of Development of Uzbekistan* (translated from the Uzbek language). Tashkent. https://cis-legislation.com/document.fwx?rgn=128505.

———.2020g. *Innovative Technologies to be Implemented Widely in the Construction Industry* (translated from the Uzbek language). Tashkent. https://president.uz/en/lists/view/3260.

———.2020h. *Resolution of the President About Additional Measures for Further Enhancement of Education System and Education* (translated from the Uzbek language). Tashkent. https://cis-legislation.com/document.fwx?rgn=128640.

Graham, P. 2021. Don't Stress About a Job Interview with Bodyswaps' New Simulator. *VR Focus*. 27 May. https://www.vrfocus.com/2021/05/dont-stress-about-a-job-interview-with-bodyswaps-new-simulator/.

Gulf News. 2021. Dubai: Sheikh Mohammed bin Rashid Congratulates Uzbek President on Independence Day. 1 September. https://gulfnews.com/uae/government/dubai--sheikh-mohammed-bin-rashid-congratulates-uzbek-president-on-independence-day-1.1630521599797.

Hanaphy, P. 2020. Adidas Reveals Futurecraft Strung, the "Ultimate" 3 Printed Running Shoe. *3D Printing Industry*. 9 October. https://3dprintingindustry.com/news/adidas-reveals-futurecraft-strung-the-ultimate-3d-printed-running-shoe-177073/.

HR Technologist. 2019. How AR and VR are Revolutionizing Soft Skills Training in 2019. 22 April. https://www.spiceworks.com/hr/learning-development/articles/how-ar-and-vr-are-revolutionizing-soft-skills-training-in-2019/

IBM Security. 2021. *Cost of a Data Breach Report 2021*. https://www.ibm.com/downloads/cas/OJDVQGRY.

IMDA. Construction and Facilities Management IDP. https://www.imda.gov.sg/programme-listing/smes-go-digital/industry-digital-plans/Construction-and-Facilities-Management-IDP.

———.Industry Digital Plans. https://www.imda.gov.sg/programme-listing/smes-go-digital/industry-digital-plans.

———.SMEs Go Digital. https://www.imda.gov.sg/programme-listing/smes-go-digital.

Institute for Workers and Trade Unions. 2020. *Automation and Its Impact on Employment in the Garment Sector of Vietnam*. http://library.fes.de/pdf-files/bueros/vietnam/17331.pdf.

Institute of Technical Education (ITE). Industry Training Schemes. https://www.ite.edu.sg/employers/industry-training-schemes/certified-on-the-job-training-centre.

International Labour Organization (ILO). 2016. *ASEAN in Transformation Textiles, Clothing and Footwear: Refashioning the Future*. Geneva. https://www.ilo.org/wcmsp5/groups/public/---ed_dialogue/---act_emp/documents/publication/wcms_579560.pdf.

———.2021. *Women and the World of Work in Uzbekistan. Towards Gender Equality and Decent Work for All*. Geneva. https://www.ilo.org/moscow/information-resources/publications/WCMS_776476/lang--en/index.htm.

INTI International University & Colleges. 2019. Industry Attachment Programme Redefines Teaching at INTI. https://newinti.edu.my/industry-attachment-programme-redefines-teaching-at-inti/.

ITU News. 2018. How Pakistan is Promoting Women and Girls in ICT. 7 March. https://news.itu.int/how-pakistan-is-promoting-women-and-girls-in-ict/.

iWeb. 2020. Back to School: Robotics, Coding Curriculum Pushed Back. 15 January. https://www.itweb.co.za/content/KzQenMj8jrzvZd2r.

K-MOOC. Courses. http://www.kmooc.kr/courses.

Lake, K. 2018. Stitch Fix's CEO on Selling Personal Style to the Mass Market. Harvard Business Review. May. https://hbr.org/2018/05/stitch-fixs-ceo-on-selling-personal-style-to-the-mass-market.

Market Screener. 2020. Government of Republic of Uzbekistan: A New System for Working with Women and Youth Defined. 10 August. https://www.marketscreener.com/news/latest/Government-of-Republic-of-Uzbekistan-nbsp-A-new-system-for-working-with-women-and-youth-defined--31507529/.

McKinsey & Company. 2012. *Delivery 2.0: The New Challenge for Governments.* https://www.mckinsey.com/industries/public-sector/our-insights/delivery-20-the-new-challenge-for-governments.

————. 2018. *Is Apparel Manufacturing Coming Home?* https://www.mckinsey.com/~/media/mckinsey/industries/retail/our%20insights/is%20apparel%20manufacturing%20coming%20home/is-apparel-manufacturing-coming-home_vf.ashx.

————. 2020. *How COVID-19 Has Pushed Companies Over the Technology Tipping Point—and Transformed Business Forever.* https://www.mckinsey.com/business-functions/strategy-and-corporate-finance/our-insights/how-covid-19-has-pushed-companies-over-the-technology-tipping-point-and-transformed-business-forever.

ME Corp Malaysia. SME's Skills Upgrading Programme. https://www.fmm.org.my/images/articles/branches/Negeri_Sembilan/SME%20PROG.pdf.

Ministry of Trade and Industry. ITMs Construction. https://www.mti.gov.sg/ITMs/Built-Environment/Construction.

————. ITMs Overview. https://www.mti.gov.sg/ITMs/Overview.

New Zealand Qualifications Authority. Micro-credentials. https://www.nzqa.govt.nz/providers-partners/approval-accreditation-and-registration/micro-credentials/

Nicholls-Lee, D. 2021. The Dutch AI startup making online clothes shopping more inclusive. Dutch News.nl. 6 January https://www.dutchnews.nl/features/2021/01/the-dutch-ai-startup-making-online-clothes-shopping-more-inclusive/.

Organisation for Economic Co-operation and Development (OECD). 2017. *Boosting SME Internationalisation in Uzbekistan Through Better Export Promotion Policies.* Paris. https://www.oecd.org/eurasia/competitiveness-programme/central-asia/Uzbekistan_Peer_review_note_dec2017_final.pdf.

————. 2019. *Sustainable Infrastructure for Low-Carbon Development in Central Asia and the Caucasus: Hotspot Analysis and Needs Assessment.* Paris. https://www.oecd-ilibrary.org/sites/5fd38a3d-en/index.html?itemId=/content/component/5fd38a3d-en.

Punjab Skills Development Fund. Training Service Providers. https://www.psdf.org.pk/tsp/industry/

Qasim, M. 2021. $150 Million Revenue brought by "Pakistani Freelancers" in a Year. https://startuppakistan.com. pk/150-million-revenue-brought-by-pakistani-freelancers-in-a-year/.

Review.Uz. 2021. State Program "Year of Support for Youth and Health Promotion" (Государственная программа «Год поддержки молодёжи и укрепления здоровья населения»). 3 February. https://review.uz/post/gosudarstvennaya-programma-god-podderjki-molodyoji-i-ukrepleniya-zdorovya-naseleniya.

Saeb, P. 2017. Virtual Reality Potential for Training Contact Centre Agents. contact-centres.com. 3 January. https://contact-centres.com/virtual-reality-potential-training-contact-centre-agents/.

Schmelzer, R. 2019. AI Applications in Education. Forbes. 12 July. https://www.forbes.com/sites/cognitiveworld/2019/07/12/ai-applications-in-education/?sh=13bf1b6b62a3.

Sierra Wireless. Intellinium Selects Sierra Wireless' Device-to-Cloud IOT Solution for Industry's First Smart Safety Shoe to Protect Workers. https://www.sierrawireless.com/company/newsroom/pressreleases/2018/02/intellinium-selects-sierra-wireless-device-to-cloud-iot-solution-for-smart-safety-shoe/.

Skillnet Ireland. Our Support for Businesses. https://www.skillnetireland.ie/about/our-support-for-smes/.

State Committee of the Republic of Uzbekistan on Statistics. Construction. https://stat.uz/en/official-statistics/construction.

———. Industry. https://stat.uz/en/official-statistics/industry.

———. Labor Market. https://stat.uz/en/official-statistics/labor-market.

Techwire Asia. 2021. The Case for Digital Twins in Construction and Real Estate. 25 August. https://techwireasia.com/2021/08/the-case-for-digital-twins-in-construction-and-real-estate/.

The Phnom Penh Post. 2021. Factory Literacy Programme Expands. 29 March. https://www.phnompenhpost.com/national/factory-literacy-programme-expands.

The Straits Times. 2019b. New $2.2m Construction-Safety School Will Use VR to Let Workers See How Dangerous Worksites Are. 14 June. https://www.straitstimes.com/business/new-22mil-construction-safety-school-will-use-virtual-reality-to-let-workers-see-how.

———. 2019a. 3D-printed Features to Debut in Tengah, Bidadari Estates. 16 September. https://www.straitstimes.com/singapore/housing/3d-printed-features-to-debut-in-tengah-bidadari-estates.

The Tashkent Times. 2017. Uzbekistan's Development Strategy for 2017–2021 Has Been Adopted Following Public Consultation. 8 February. http://tashkenttimes.uz/national/541-uzbekistan-s-development-strategy-for-2017-2021-has-been-adopted-following-discussion

Triax. 2021. Triax Launches Simplified IOT Solution Providing Preventative Safety, Productivity Tools; Furthers Digital Transformation on Worksites. https://www.triaxtec.com/news/press-release/triax-launches-simplified-iot-solution-furthers-digital-transformation-on-worksites/.

Tukatech. 2021. Levi's Largest Pakistan Knit Supplier Expands Capacity with TUKA Cutter. 14 May. https://tukatech.com/combined-fabrics-third-fabric-cutter/.

Tursunov, N. N. 2017. *Study on Development Strategy for Uzbekistan Clothing Industry*. https://core.ac.uk/download/pdf/213852882.pdf.

United Nations Industrial Development Organization (UNIDO). 2021. *Innovation and Industrialization for SDG9 in Uzbekistan. Vienna*. https://tii.unido.org/sites/default/files/publications/Discussion%20paper%20on%20SDG%209%20in%20Uzbekistan.pdf.

UZ Daily. 2017. Uzbekistan Adopts Measures to Accelerate Development of Textile, Sewing and Knitting Industries. 15 December. https://www.uzdaily.uz/en/post/41969.

———. 2019a. Large-Scale "One Million Uzbek Coders" Project Launched in Uzbekistan. 21 November. https://www.uzdaily.uz/en/post/53182.

———. 2019b. Concept of the Development Strategy of Uzbekistan-2035 put for discussion. https://www.uzdaily.uz/en/post/49727.

———. 2020. Business Clinic Will Support Business During and After the Pandemic. 3 June. https://www.uzdaily.uz/en/post/57480.

———. 2021a. The Volume of Construction Work Performed in Uzbekistan Makes Up 5.86 Trillion Soums. 22 February. https://www.uzdaily.uz/en/post/63794.

———. 2021b. Training Center TECHNONICOL to Give Impetus to the Development of Energy Efficient Technologies in Uzbekistan. 3 June. https://www.uzdaily.uz/en/post/65828.

———. 2021c. Uzbek Leader Holds Meeting to Analyze the Performance of Tasks in the Field of Youth Policy. 13 April. https://www.uzdaily.uz/en/post/64820.

———. 2021d. A New Joint Project "Future Skills Uzbekistan" Has Been Launched. 21 November. http://www.uzdaily.uz/en/post/62947.

Virtusize. About Us. https://www.virtusize.com/

Waya. 2019. How IOT Is Transforming the Garment Industry. 10 April. https://waya.media/how-iot-is-transforming-the-garment-industry/.

World Bank. 2020. Uzbekistan Public Expenditure Review. Washington, DC. https://www.worldbank.org/en/country/uzbekistan/publication/per.

Xin Hua. 2021. Uzbekistan Increases Exports of Textile Products in 2020. 27 January. http://www.xinhuanet.com/english/2021-01/27/c_139702034.htm.

Your story. 2020. This Edtech Startup is Using Virtual Reality-Based Content to Have Real-Life Impact on Students. https://yourstory.com/2020/05/startup-bharat-edtech-startup-fotonvr-vr-schools/amp.

zSpace. Beyond STEM: Building Soft Skills with Augmented and Virtual Reality. https://zspace.com/blog/going-beyond-stem-building-soft-skills-with-augmented-and-virtual-reality.